Hybrid Vehicle Basics

Presented By:
Automotive Aftermarket Training, Inc.
PO Box 836
Forestdale, MA 02644
877-488-5472

This publication, and any associated course(s), is/are intended to be an introduction only to hybrid vehicles, their design, and their unique service and safety requirements. Course content is presented for information only. Hybrid vehicles have unavoidable hazards including high voltage electric shock, high temperature liquids, and hot components.

Automotive Aftermarket Training, Inc. (AAT), their employees, agents, and suppliers, are not responsible for any consequential injuries or damages from your decision to perform any and all maintenance and repair procedures to a hybrid vehicle.

Caution:

Hybrid systems have high-voltage circuits that can carry up to 650 volts. Risk of electrical shock and electrocution!

Do:

Wear lineman's gloves whenever working near potential high voltage.

Be sure hybrid system is OFF and safe before performing any service, maintenance, or repair.

Be sure to test to ensure that high voltage is not present before performing any service, maintenance, or repair.

Be sure vehicle is in PARK with parking brake applied before exiting vehicle.

Disconnect the 12-volt battery if a hybrid vehicle will not be used for more than three weeks.

Tow/transport hybrid vehicles so that the drive wheels do not turn (may generate high voltage).

Do Not:

Run hybrid vehicle out of gas. A fully discharged high voltage battery may be impossible to recharge.

Charge the 12-volt auxiliary battery in a Prius with a conventional high-rate battery charger.

Leave ignition ON and not in Ready Mode for extended time. This will allow the 12-volt auxiliary battery to discharge.

1

Contents

(This page intentionally left blank)

What is a Hybrid?

Learning objectives for this Section:

- ✓ You will be able to define a hybrid vehicle.
- ✓ You will be able to list hybrid vehicle advantages and disadvantages.
- ✓ You will be able to describe the types of hybrids and typical hybrid operating modes.

Introduction

A hybrid vehicle has more than one power source that drives the vehicle. The most common configuration today combines an electric motor with an internal combustion (IC) engine. Shortly you will see how power to the wheels can come from the IC engine, an electric motor, or a combination of both. Under some conditions, the vehicle can be moving under electric power only and the IC engine may not be running.

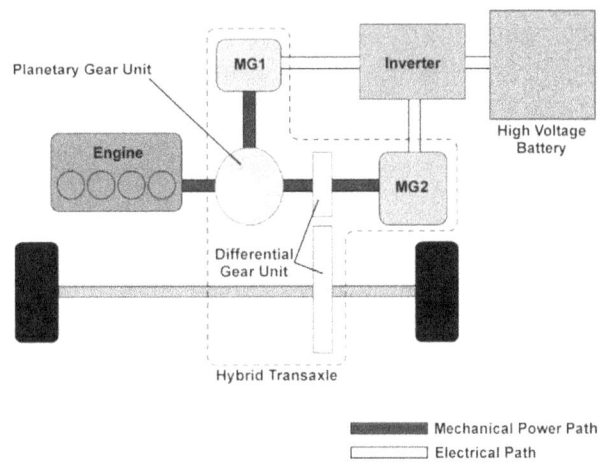

Energy to move the vehicle is stored two ways: gasoline in the fuel tank stores chemical energy, and a high-voltage battery stores electrical energy. Gasoline is a high-energy fuel and contains about 130,000 BTUs of energy per gallon. That's about 20,800 BTU/lb or 10 Watt-hrs/gram. A typical lead-acid 12-volt automotive battery stores only 0.3% as much energy per pound, and even the best batteries currently available store only 1.5% as much energy per pound compared to gasoline. The high energy content in a gallon of gasoline is one reason why it has not easy to find alternative energy sources.

Advantages of Hybrids

Like any other complex object with difficult engineering and design challenges, there are strong and weak points, advantages and disadvantages. Here are a few of each:

- IC engines can be very efficient when running in a constant-speed-constant-load mode of operation. Hybrid control systems make the IC engine run efficiently more of the time, and avoid inefficient modes of operation (idle, full throttle, high rpm, etc.).

- Usually the IC engine can be completely OFF at stop to provide zero emissions and zero fuel consumption. The engine will restart as soon as the accelerator pedal is depressed.

- Electric motors provide highest torque when starting up, such as when getting the vehicle moving. This high torque from the electric motor can remove a load from the IC engine when it is very inefficient. Sometimes, under light load and low speeds, the electric motor alone will propel the vehicle without the IC engine running.

- Regenerative braking can recover energy that can be stored as electrical energy in a battery and can be reused. Conventional brakes convert all of the kinetic energy of motion to heat, which is dissipated to the ambient air. Regenerative braking converts this kinetic energy to electrical energy which is used to recharge the onboard high-voltage battery instead of being lost as heat.

- Conventional disk and drum brakes are used. The brakes are used less, so they last longer, and brake repairs are done in the same way as on non-hybrid vehicles.

- Hybrids have very clean emissions ratings. Pollution and carbon dioxide (CO_2) emissions are reduced, and fuel economy is improved.

Disadvantages of Hybrids

- Complexity of mechanical and electronic systems and powertrain controls.

- Complexity of mechanical and electronic systems and controls for the brakes – the system must coordinate regenerative braking with conventional friction braking and stop the vehicle in a normal, predictable way.

- Complex and expensive manufacturing that requires much energy input.

- Use of rare and difficult to obtain natural resources, such as large rare earth permanent magnets in brushless electric motors.

Types of Hybrids

In a **Series Hybrid** system, electricity from a generator and electricity stored in a battery power an electric motor which drives the vehicle. An internal combustion engine drives the generator. The wheels are driven only by the electric motor and not by the IC engine. So a Series Hybrid is an electrically powered car with an onboard engine-driven generator.

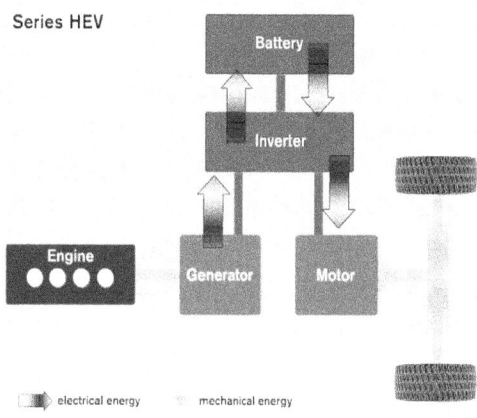

A **Parallel Hybrid** system can use both the internal combustion engine and the generator to directly drive the wheels. Very sophisticated electronic controls manage the amount of power contributed from the engine and electric motor. The driver is usually unaware of the constantly shifting contribution from each power source. The electric motor can also function as an engine-driven generator to recharge the storage battery. Hybrid vehicles are rated by what proportion of the power is contributed by the electric motor.

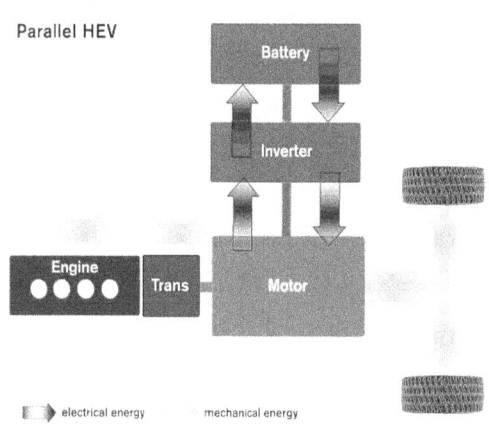

Another way to rate or categorize hybrids is by the amount of power contributed by the alternative power source:

A **Mild Hybrid** relies mostly on the IC engine to move the vehicle. The electric assist to the wheels is minimal, and "engine-off-at-stop" capability is usually the major benefit. The Chevy Malibu Hybrid is a mild hybrid.

A **Full Hybrid** vehicle moves with power from the IC engine, the electric motor, or a combination of both. The actual contribution from each power source depends on vehicle design and the mode of operation. The engine supplies power to the drive wheels and also drives a generator to supply power to both the electric motor and to the high voltage battery. All models of the Toyota Prius are full hybrids.

Typical operating modes

Below are simple block diagrams showing typical operating modes for the most common type of hybrid vehicle powertrain. This is how Toyota, Ford, and Nissan hybrids are configured, and they all work in a similar way. Toyota calls their system the Toyota Hybrid System (THS). A newer version is called Toyota Hybrid System II (THS II). Nissan licenses Toyota technology and uses some Toyota components in their Altima Hybrid. Ford has its own transaxle manufactured by Aisin and it works similar to the Toyota system.

In THS, THS II, and similar systems, there are two brushless alternating current (AC) motor/generators, typically called MG1 and MG2. The block diagrams show when power is contributed by the IC engine and by the electric motor/generators. The diagrams also show under what conditions the high-voltage (HV) battery is discharged and when it is charged.

Motor/generator MG2 is geared directly to the front wheels through the final drive. Any time the wheels turn, MG2 turns. MG2 can be powered as a motor to drive the wheels, or the turning wheels can drive MG2 as a generator.

Motor/generator MG1 can act as a starting motor for the IC engine. When the IC engine is running, it can drive MG1 as a generator to provide voltage and current to run MG2, and also to recharge the high-voltage battery.

All three power components - the IC engine, MG1, and MG2, are mechanically linked with a simple planetary gear set. You will see this in the next Section.

Stopped

When the hybrid system is ON, hybrid system control modules are powered up and the vehicle is ready to move.

The IC engine may or may not run under these conditions and may start at any time. This is an important safety concern – more on this later!

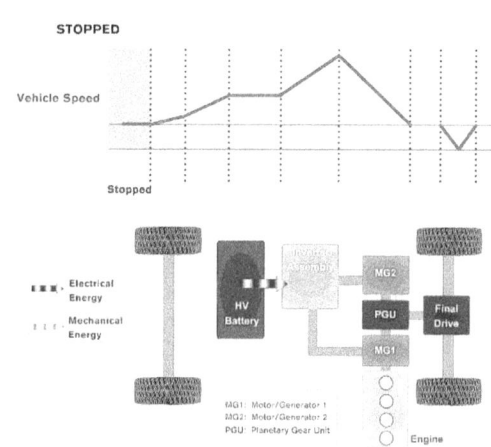

Starting to move

When the vehicle begins to move off slowly, it is usually driven only by electric motor MG2. The IC engine will not start unless conditions require it to run, such as low HV battery voltage or during engine warm up.

Power to drive MG2 is provided by the HV battery.

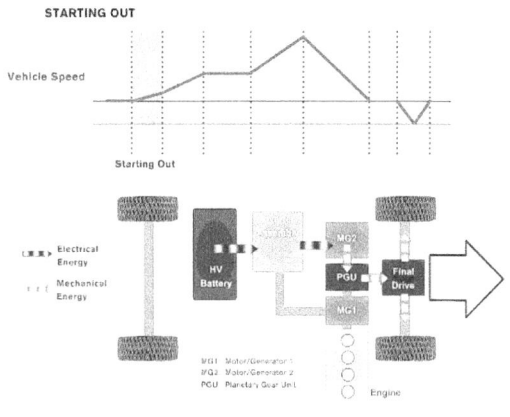

IC engine startup

When road speed and requested acceleration exceed predetermined values, the system starts the IC engine in order to supply additional power to the wheels. To start the engine, power from the HV battery turns MG1, which acts like a cranking motor.

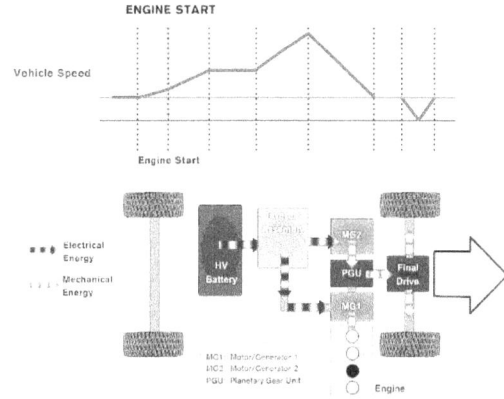

Acceleration

Power to accelerate is contributed by both the IC engine and MG2. The IC engine also turns MG1 which acts as a generator to supply electrical power to MG2. The amount of power contributed by the engine and by MG2 is determined and controlled by the hybrid control system.

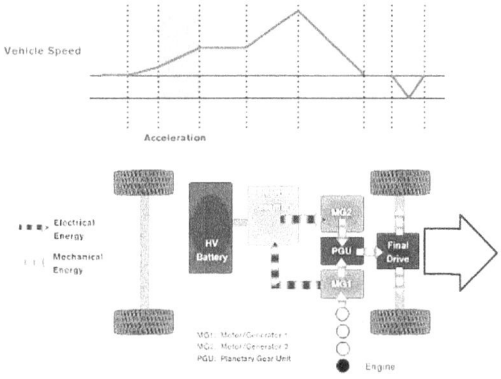

Cruise

During part-throttle cruise, both the IC engine and MG2 typically drive the vehicle. The engine also turns MG1 which acts as a generator to supply electrical power to MG2 and to recharge the HV battery.

Changes in terrain and accelerator pedal position may cause the engine to stop and restart.

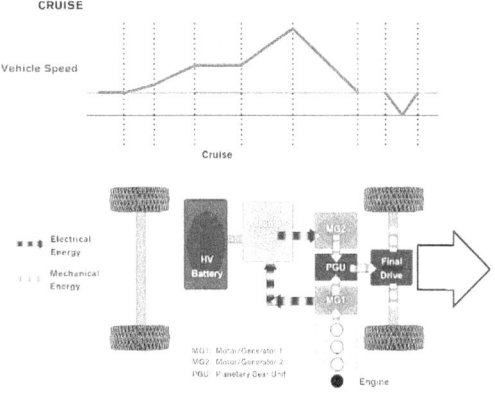

Hard acceleration

During hard acceleration, MG2 and the IC engine both drive the vehicle. Both MG1 and the HV battery supply extra electrical power to MG2 for maximum performance.

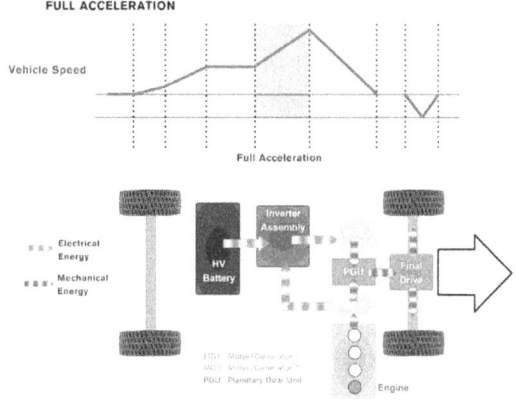

Deceleration and braking

During deceleration and braking, the IC engine typically does not run. The vehicle is slowed by regenerative braking and also by conventional mechanical brakes.

Regenerative braking occurs when MG2 is turned as a generator by the rotation of the vehicle's wheels and recharges the HV battery. Torque required to turn this high-voltage generator slows the vehicle.

Reverse

With the THS/THS II type of hybrid powertrain, there is no reverse gear in the transmission. The only way to drive in reverse is electrically. Electric motor MG2 is powered to run in the reverse direction to cause the vehicle to back up.

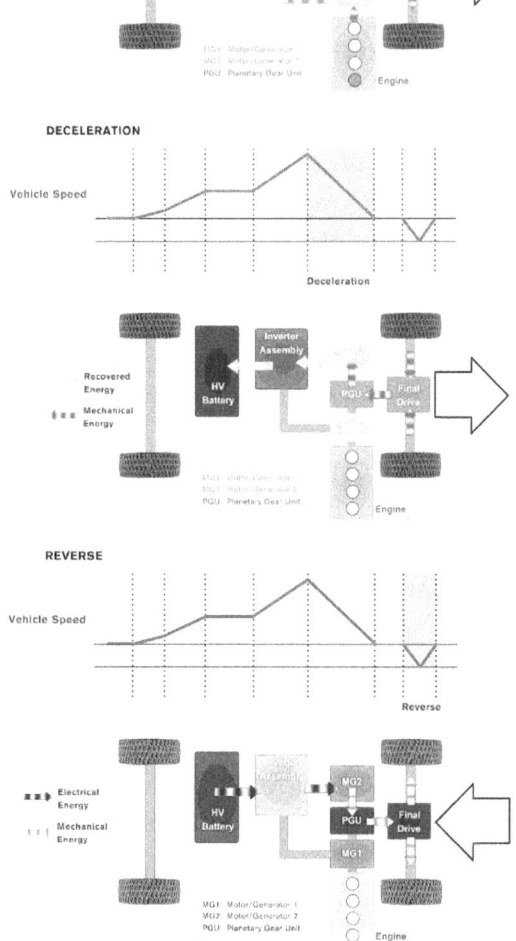

Hybrid sales in the US

As the table below illustrates, hybrid vehicle sales in the US peaked in 2007 at 352,000 units. Production years 2008 and 2009 have shown a minor reduction in vehicle production (313,000 and 292,000, respectively) due to an economic recession.

Out of a total of approximately 606,000 total hybrid vehicles sold in 2008 and 2009:

- Toyota hybrid vehicles: 400,000 (66%)
- Honda hybrid vehicles: 67,000 (11%)
- Ford hybrid vehicles: 55,000 (9%)
- Lexus hybrid vehicles: 38,000 (6%)

Total U.S. sales for the Prius (all generations) have exceeded 1,000,000 vehicles, so it is more than likely you will be seeing a Prius in your shop. Other vehicles that use the THS/THS II system include the Highlander, Camry and Nissan Altima. And the Ford hybrid system is very similar to the Toyota system.

Honda has also sold over 230,000 hybrid vehicles in the US since 1999. The Civic is the most popular with 191,000 vehicles sold.

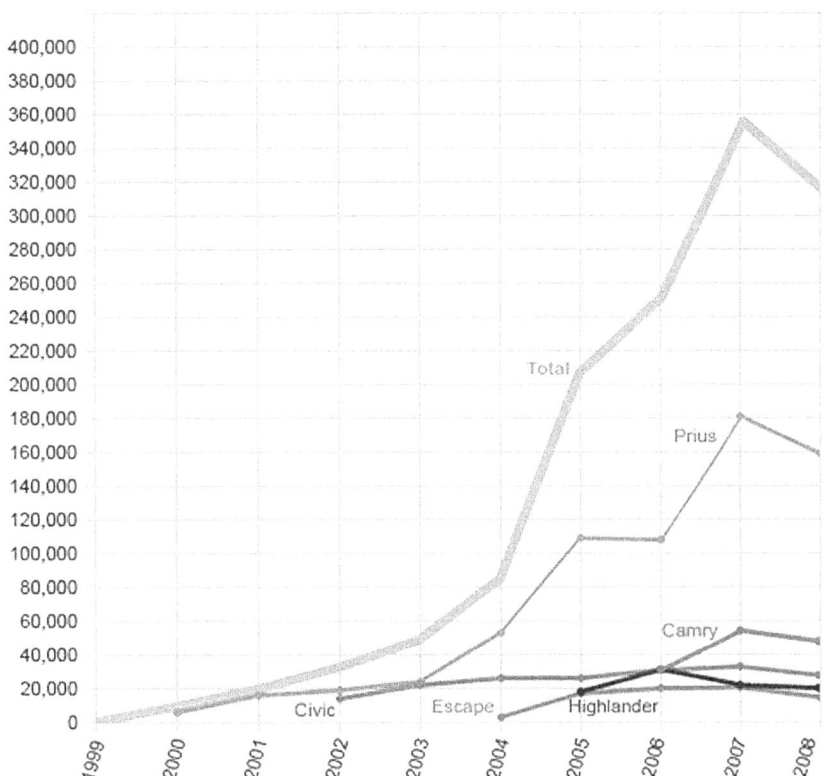

Hybrids you may see in your shop

Toyota and Lexus	Model Years Avail.	Hybrid Type
Prius 1st gen (NLA*)	2001-2003	full
Prius 2nd gen (NLA*)	2004-2009	full
Prius 3rd gen	2010>	full
Highlander Hybrid	2006>	full
Camry Hybrid	2007>	full
Lexus LS 600h	2006>	full
Lexus GS 450h	2007>	full
Lexus RX400h/450h	2005>	full
Lexus HS 250h	2010>	full
Nissan	**Model Years Avail.**	**Hybrid Type**
Altima Hybrid	2007>	full
Ford	**Model Years Avail.**	**Hybrid Type**
Escape Hybrid	2005>	full
Mercury Mariner Hybrid	2005>	full
Fusion Hybrid	2009>	full
Mercury Milan Hybrid	2009>	full
Honda	**Model Years Avail.**	**Hybrid Type**
Insight 1st gen (NLA*)	2000-2006	full
Insight 2nd gen	2010>	full
Civic Hybrid	2003>	full
Accord Hybrid (NLA*)	2005-2007	full
General Motors	**Model Years Avail.**	**Hybrid Type**
Saturn Vue Green Line Hybrid (NLA*)	2007-2009	mild
Saturn Aura Hybrid (NLA*)	2008-2009	mild
Chevrolet Malibu Hybrid (NLA*)	2008-2009	mild
Chevrolet Tahoe Hybrid	2008>	full (two-mode)
GMC Yukon Hybrid	2008>	full (two-mode)
Chevrolet Silverado Hybrid	2004>	full (two-mode)
Cadillac Escalade Hybrid	2009>	full (two-mode)
GMC Sierra Hybrid	2004>	full (two-mode)
BMW	**Model Years Avail.**	**Hybrid Type**
Active Hybrid X6	2010>	full (two-mode)
Active Hybrid 7	2010>	mild

(*NLA = no longer available)

In addition to the above list, several more hybrid models have been sold in the US, but the numbers are very small. Examples include the Chrysler Aspen and Dodge Durango Hybrids, which used the two-mode hybrid transmission that was co-developed by GM, BMW, and Chrysler.

Hybrid Safety

Learning objectives for this Section:

✓ You will be able to list the mechanical and electrical dangers present in hybrid vehicles.
✓ You will be able to explain correct procedures to disable hybrid HV systems.
✓ You will be able to describe how to test HV components and systems for safety prior to beginning any repair procedure.

Introduction

Hybrid vehicles are in many ways no different than conventionally powered vehicles and have many of the same hazards: rotating components, compressed springs, pinch points, and hot engine, exhaust, and cooling system parts. In addition, hybrids add the hazards of high voltage and the possibility of an unexpected engine start. This Section will describe these potential hazards and how to avoid them.

Mechanical dangers

Engine may start unexpectedly: In Toyota, Lexus, and Nissan hybrid vehicles, when the ignition switch is turned ON, a READY indicator will illuminate on the instrument panel.

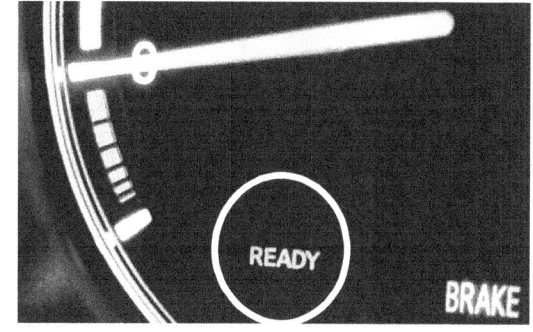

When the system is in the Ready Mode, the engine may start and stop at any time and without warning. This is most important when performing any service and when working near rotating parts. The hybrid system must be OFF before any service is performed.

Honda hybrid vehicles have an Auto Stop Mode, which is similar to the Ready Mode in Toyota hybrids.

When in Auto Stop Mode, the "Auto Stop" indicator on the instrument panel will illuminate. When in Auto Stop Mode, the engine may start at any time and without warning. The hybrid system must be OFF before any service is performed.

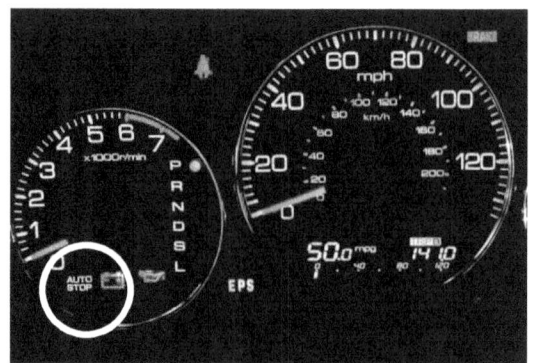

GM mild and two-mode hybrids have an Auto Stop Mode, which is similar to the Ready Mode in Toyota hybrids.

When in Auto Stop Mode, the needle indicator on the instrument panel will point to "Auto Stop." When in Auto Stop Mode, the engine may start at any time and without warning. The hybrid system must be OFF before any service is performed.

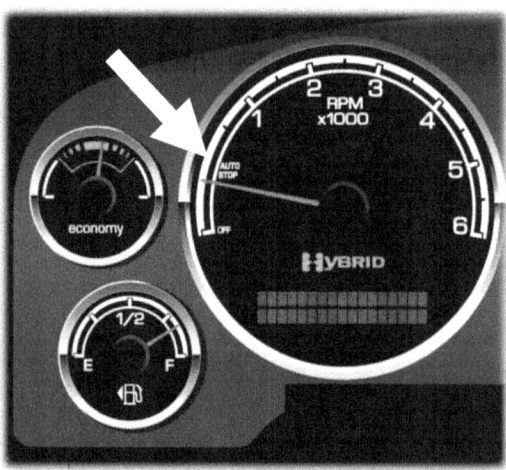

CAUTION!

When the hybrid system is ON and in the Ready Mode or Auto Stop Mode, the IC engine may start at any time. Be sure to turn the hybrid system OFF before performing any service.

Smart Key: Vehicles with the Smart Key system can be started without inserting the key into a dashboard opening or lock cylinder. If the vehicle is equipped with Smart Key, remove the key from the vehicle and safely store it at least 15 feet away from the vehicle prior to starting any service or repair.

For additional safety, disable the Smart Key function. This graphic shows the location of the Smart Key switch on the 2004-2009 Prius.

CAUTION!

Remove and safely store a Smart Key at least 15 feet away from the vehicle before performing any service.

Hot coolant: The engine cooling system in the 2004-2009 Prius has a unique container located in the left front fender that stores hot coolant for an extended period. The tank can store coolant as hot as 176° F (80° C) for up to three days. When working on the engine cooling system on a 2004-2009 Prius, be prepared for this hot coolant.

CAUTION!

The coolant heat storage tank in the 2004-2009 Prius stores coolant as hot as 176 degrees F for up to three days. Hot coolant can scald. Use extreme caution.

High voltage dangers

Shock hazard: Hybrid vehicles have high voltage circuits that are designed with technician and occupant safety in mind. However, there are always inherent dangers when working around high voltage electricity. Treat HV components with respect, just like house wiring.

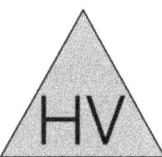

Hybrid vehicles have high voltage electrical circuits, capacitors, electric motor / generators, wiring, and other potentially dangerous high voltage components. Voltages can be as high as 650 volts AC or DC, depending upon the vehicle manufacturer. Voltages greater than 25 volts (rms) AC or 60 volts DC can cause sufficient current to pass through the body to be considered dangerous. Even a very small current through the body can cause injury or death, and currents greater than 100-200 mA can be fatal.

Current		Reaction
Safe current values	1 mA or less	Causes no sensation – not felt
	1 mA to 8 mA	Sensation of shock, not painful, individual can let go at will because muscular control is not lost
Unsafe current values	8 mA to 15 mA	Painful shock, individual can let go at will because muscular control is not lost
	15 mA to 20 mA	Painful shock, control of adjacent muscles is lost, victim can not let go
	50 mA to 100 mA	Ventricular fibrillation, a heart condition that can result in death is possible at this current level
	100 mA to 200 mA	Ventricular fibrillation occurs
	200 mA and over	Severe burns, severe muscular contractions, so severe that chest muscles clamp the heart and stop it for the duration of the shock (This prevents ventricular fibrillation)

CAUTION!

Hybrid vehicles have voltages as high as 650 volts AC or DC. Use recommended safety equipment and verify that high voltage is not present before touching hybrid system components.

Identifying high-voltage wiring

Hybrid vehicles usually locate the HV battery in the rear and the electronic controls up front near the engine. Heavy high voltage high current capacity cables run between the two. These cables run in protected channels under the floor pan or through the passenger compartment. Both positive and negative HV wiring is insulated from the vehicle body.

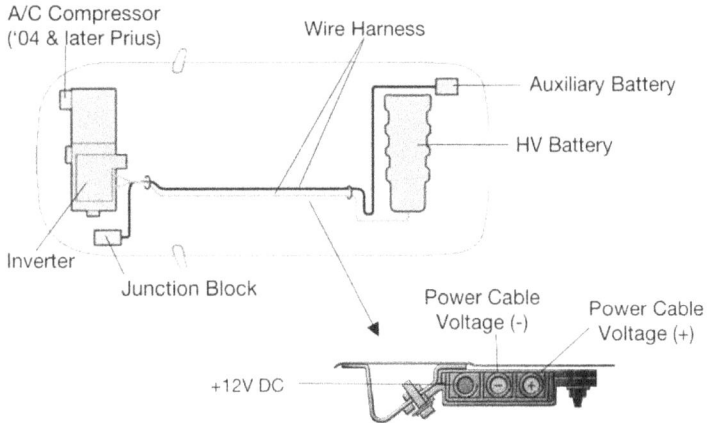

The industry has standardized high-voltage wiring color for safety. All high-voltage wiring, connectors, and insulation are colored orange. Some cables are protected by flexible jackets or rigid plastic covers.

Remember, these high-voltage circuits may carry up to 650 volts. This course will describe manufacturer's procedures for isolating and opening high-voltage circuits prior to beginning any service or repair that includes hybrid system components.

The Toyota Camry and Highlander hybrids use a 42-volt electric power steering system. Wiring to the steering system electric motor is yellow. Do not confuse this high-voltage wiring with 12-volt yellow wiring to Supplemental Restraint System (SRS) components such as airbags, impact sensors, and controllers.

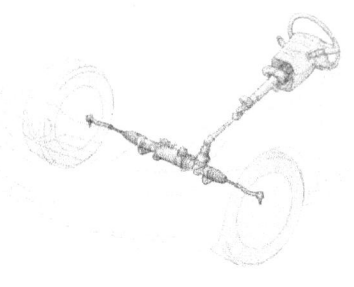

GM mild hybrids with the Belt Alternator Starter (BAS) use a 36-48 volt hybrid system. The HV wiring in these vehicles is colored blue.

High voltage (HV) batteries

Toyota HV batteries: Toyota HV batteries are assembled from a large number of individual nickel metal hydride (NiMH) cells. Each cell has a rigid plastic case. Each cell generates 1.2 volts. Six of these cells are connected in series to make a long and narrow 7.2 volt battery module.

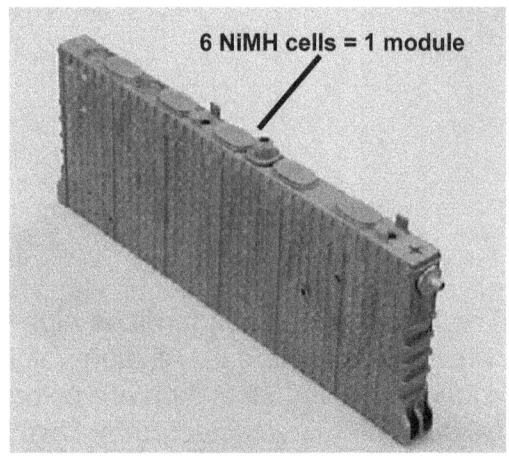

6 NiMH cells = 1 module

The battery modules are electrically connected in series. This means that they are connected positive-to-negative. This is just like placing batteries in a flashlight. Placing batteries end-to-end causes their voltage to add together.

The 7.2 volts produced by each HV battery module add together to produce the high voltage output. For example, the 2004-2009 Prius HV battery has 28 modules (28 modules x 7.2 volts = 201.6 volts). The battery module terminals are threaded studs. Copper bus bars are fastened to the module terminals with threaded nuts.

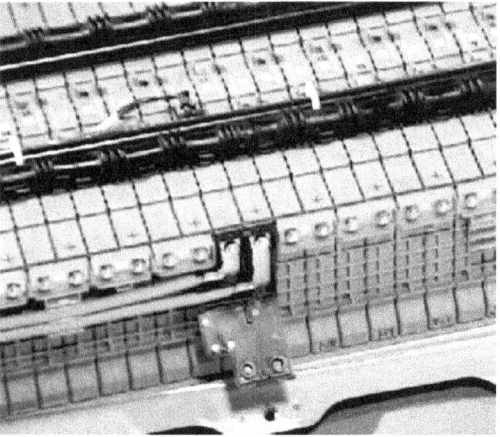

The number of modules and voltage produced by the HV battery varies by hybrid model, but all of these batteries generate voltages over 200 volts. This voltage is high enough to create a shock hazard.

SERVICE HINT: 	Toyota has a reward program for recycling HV batteries and pays a "bounty" to dealers for returned batteries. See your local dealer if HV battery disposal is necessary.

High voltage battery locations

Toyota Prius: HV batteries are typically located in the vehicle rear behind the rear seat or under the rear floor. HV batteries are usually protected by formed sheet metal housings. The HV battery in the 2001-2003 Prius is rated at 273.6 volts. The HV battery in the 2004-2010 Prius is rated at 201.6 volts.

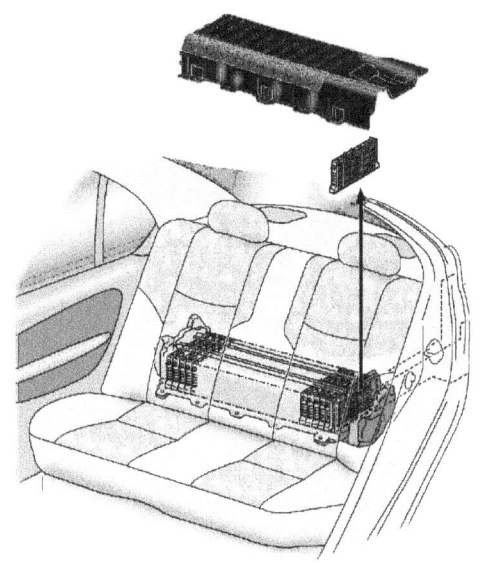

Toyota Camry Hybrid: Located behind the rear seat. HV battery supplies 244.8 volts.

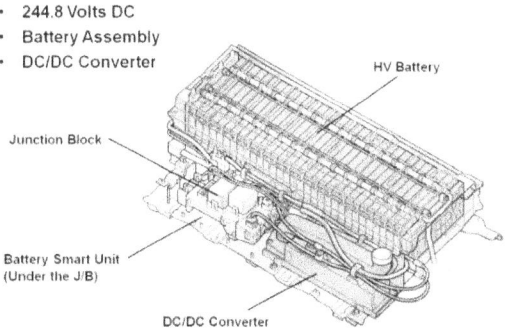

- 244.8 Volts DC
- Battery Assembly
- DC/DC Converter

HV Battery

Junction Block

Battery Smart Unit
(Under the J/B)

DC/DC Converter

Toyota Highlander Hybrid: HV battery is located under the rear seat. The battery is divided into three sections for space considerations and produces 288 volts.

Nissan Altima Hybrid: Uses the same 244.8-volt HV battery pack as the Toyota Camry Hybrid. The HV battery is also in the same location, behind the rear seat.

Ford Fusion/Mercury Milan Hybrids: HV battery is located behind the rear seat. This HV battery produces 275 volts.

Ford Escape/Mercury Mariner Hybrids: HV battery is under the rear cargo floor. This HV battery is rated at 300 volts.

Honda Hybrids: This is a typical Honda HV battery assembly. NiMH cells are cylinder-shaped like D-cell flashlight batteries.

2000-2006 Insight: 158 volts

2010> Insight: 100.8 volts

Civic Hybrid: 144 volts

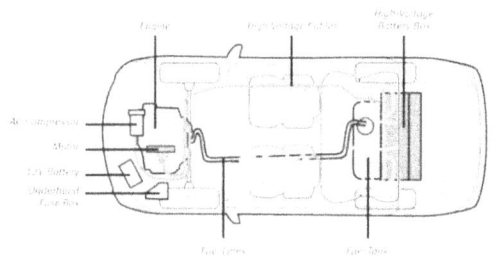

HV Battery Module   Single HV Battery "Stick"   Single HV Cell   D-Cell Battery

Honda HV batteries are located behind the rear seat.

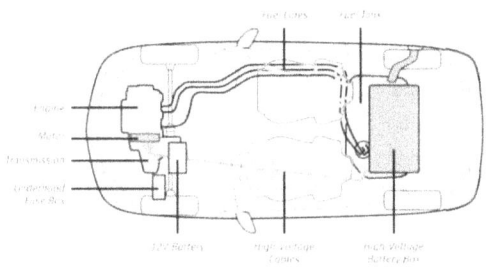

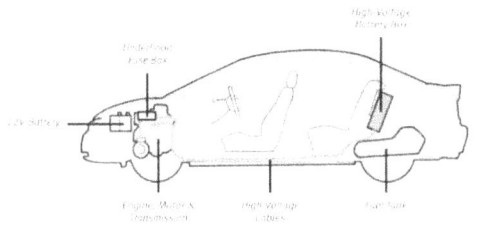

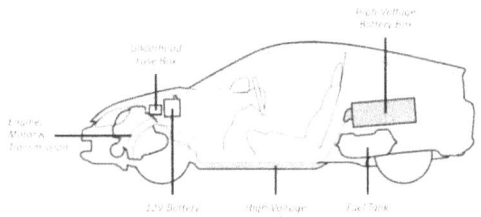

Honda 2000-2006 Insight

Honda Civic Hybrid

GM Malibu/Aura/Vue mild hybrids: The 36-volt HV battery is located behind the rear seat.

GM Tahoe/Yukon/Silverado/Sierra/ Escalade Two-mode hybrids: The 300-volt HV battery is located under the rear seat.

Chemical dangers

Nickel metal hydride HV battery safety: The electrolyte in each nickel metal hydride (NiMH) HV battery cell is a very caustic chemical paste containing sodium hydroxide and potassium hydroxide.

This chemical is very alkaline with a value of 13.5 on the pH scale, where 1 is very acidic, 7 is neutral, and 14 is very alkaline (basic). Never touch this electrolyte if it leaks!

Neutralizers for spilled electrolyte are vinegar or a dilute boric acid solution (5.5 oz. boric acid/gal. water).

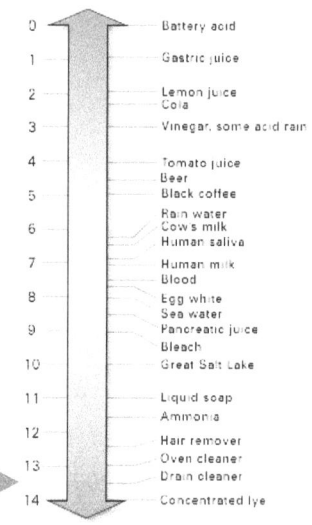

pH	
0	Battery acid
1	Gastric juice
2	Lemon juice / Cola
3	Vinegar, some acid rain
4	Tomato juice / Beer
5	Black coffee
6	Rain water / Cow's milk / Human saliva
7	Human milk / Blood
8	Egg white / Sea water
9	Pancreatic juice / Bleach
10	Great Salt Lake
11	Liquid soap / Ammonia
12	Hair remover
13	Oven cleaner / Drain cleaner
14	Concentrated lye

pH Scale

CAUTION!

Never touch NiMH electrolyte paste in the HV battery! The electrolyte is a very caustic alkaline which can cause severe burns! Flush skin with water and seek immediate medical aid.

Safety equipment

High voltage lineman's gloves: Rubber lineman's gloves should be worn whenever high voltage may be present. Wear them before and during Service Plug removal, and before you have verified that components are isolated from high voltage. The gloves must meet ANSI/ASTM Class 0 insulation specifications (1,000 volts AC).

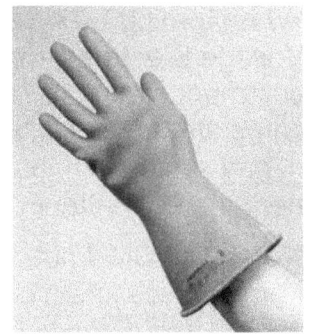

Remove all jewelry, watches, rings, chains, etc. before working on a vehicle. Store lineman's gloves in a cool dry place, and do not fold or store inside out.

CAUTION!

 ***Inspect gloves for cuts and punctures by blowing into open end and partially inflating. Hold under water if you suspect a leak. Even a pinhole can be dangerous!***

Eye protection: Use of safety glasses or goggles is a long-standing shop safety practice. Eye protection should always be worn whenever performing maintenance or repair procedures.

Fire extinguisher: Use only Class C fire extinguishers rated for electrical fires.

CAUTION!

 ***Water is not recommended for putting out electrical or battery chemical fires and should not be used as a fire extinguisher on hybrid vehicle fires.***

High voltage safety systems

System Main Relay (SMR) (Toyota THS/THS II):
When turning a Toyota hybrid system ON, the High Voltage Electronic Control Unit (HV-ECU) commands a set of System Main Relays (SMR) to complete the circuit between the HV battery and hybrid system components. The SMR is located near the HV battery.

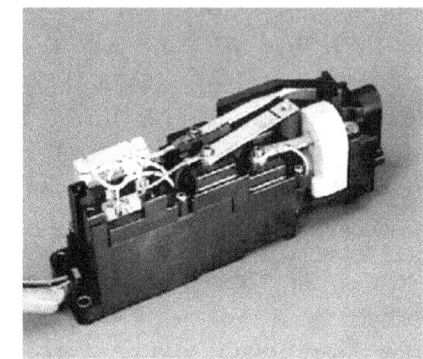

Note: The SMR not only isolates HV battery voltage from hybrid system components, but also allows for vehicle self-diagnostics.

SMR closing sequence: The SMR close in a specific sequence to reduce current inrush. The middle relay has a resistor in series.

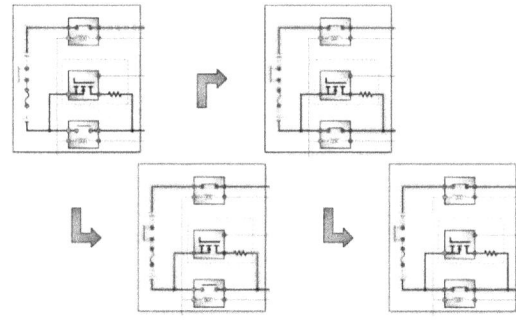

When the hybrid system turns ON, the top and middle relays close first, and current inrush is limited by the resistor in series with the middle relay. The bottom relay then closes allowing full current flow, and finally the middle relay opens since its path has been bypassed.

When you turn the hybrid system OFF, the SMR opens the circuit between the HV battery and other HV system components. When a serious fault is detected, the HV-ECU can command the SMR to open and disconnect the HV battery. This will completely shut down the hybrid system.

Note: The SMR relay closing sequence not only helps to control and protect the high-voltage system, but also helps the system check for faults.

Voltage and current leakage: Hybrid system onboard diagnostics monitor HV systems and wiring for voltage and current leakage. If unexpected voltage or current leakage is detected, the hybrid system will either run in limp-in mode or shut down completely, depending upon the severity of the fault.

This graphic shows a Prius hybrid system in the Ready Mode, with system main relays closed, connecting the HV battery to the hybrid system.

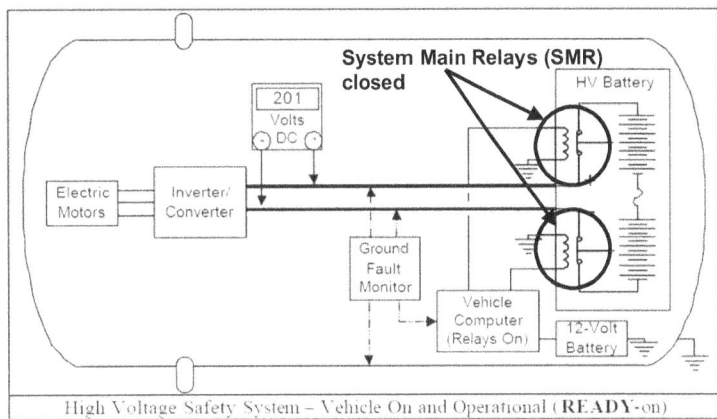

In this graphic, the system main relays are open. The HV battery is now disconnected and isolated from the hybrid system.

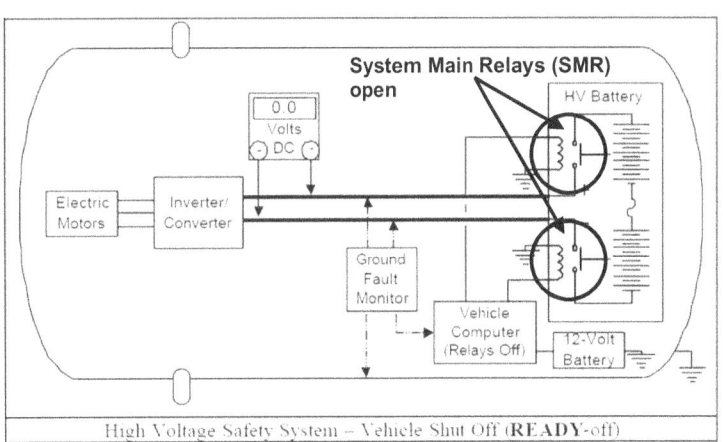

Service Plug

Most available hybrids have a manually-removable electrical connector.

Toyota calls this connector the Service Plug. When the Service Plug is installed, its contacts complete the HV battery circuit.

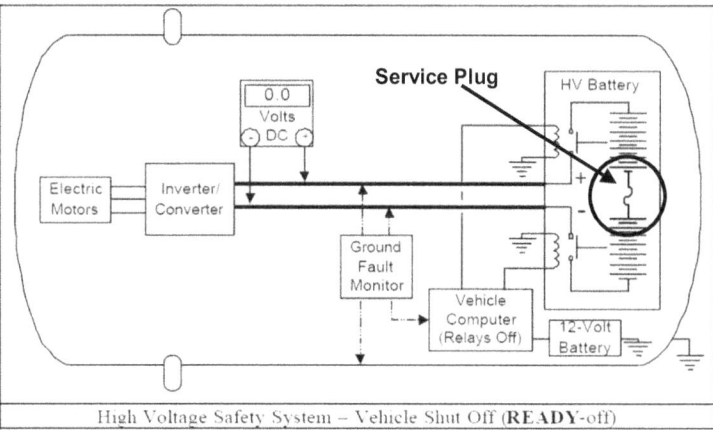

High Voltage Safety System – Vehicle Shut Off (**READY**-off)

Removing the Service Plug opens the HV battery circuit and prevents the hybrid system from turning ON.

The Service Plug must be removed in order to disable the hybrid system and to remove high voltage from hybrid system components. See **Service Plug removal and installation**, later in this Section.

The graphic above shows where a Service Plug is typically located in the circuit.

Service Plug names: Different manufacturers have given these connectors different names:

Toyota: Service Plug

Nissan: Service Disconnect Switch

Ford: High Voltage Service Disconnect Switch

Honda: Battery Module Switch

GM: Battery Disconnect Switch

Note: In this course, we will call it this connector the Service Plug.

Service Plug fuse: In addition to being an electrical connector that completes the HV battery circuit, in most applications the Service Plug also has a high-current fuse.

Service Plug interlock contacts: Some Service Plugs also have a second set of small contacts. These contacts are part of an interlock system. When the Service Plug is fully inserted and latched, the interlock contacts complete a circuit that tells the hybrid system that the Service Plug is installed. Without this signal, the hybrid system will not turn ON. This is an important and easy check!

SERVICE HINT:	The hybrid system Master Warning Light will illuminate if the interlock contacts are not fully engaged.

Service Plug locations

2001-2003 Toyota Prius: Service Plug location is in trunk on left side of HV battery housing.

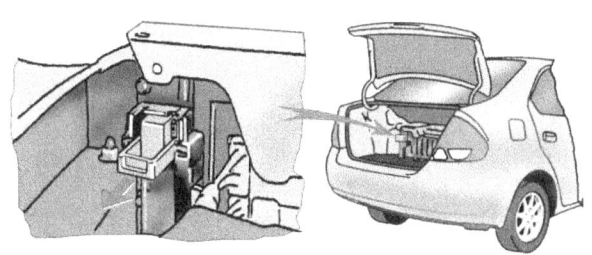

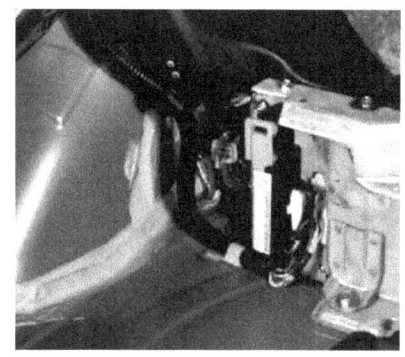

2004-2009 Toyota Prius: Service Plug location is under cargo floor on left side of HV battery housing.

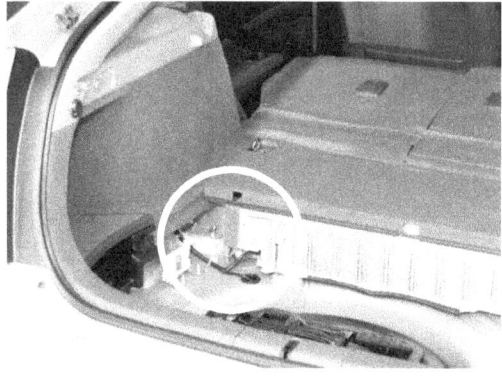

2010> Toyota Prius: Service Plug location is under cargo floor on right side of HV battery housing.

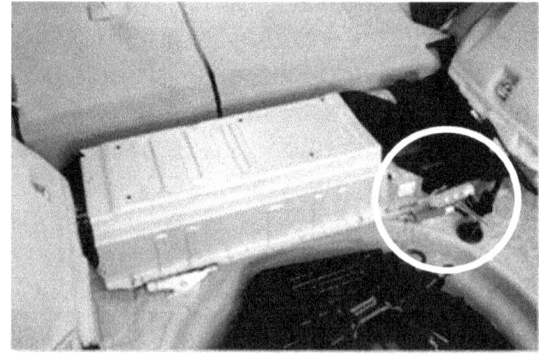

Toyota Camry Hybrid and Nissan Altima Hybrid: Service Plug location is in trunk on right side of HV battery housing.

Toyota Highlander Hybrid/Lexus RX400h /RX450h: Service Plug location is under rear seat on left side.

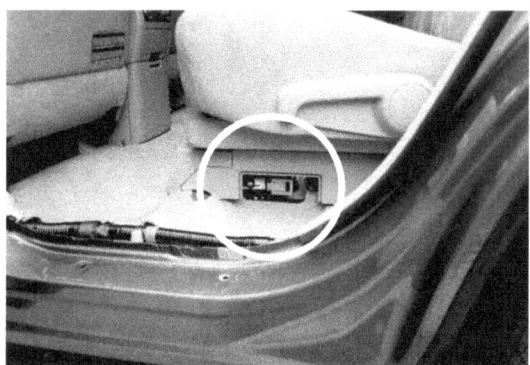

Ford Escape/Mercury Mariner Hybrid: Service Plug location is under cargo floor on right side (vehicle also has two inertia switches which can interrupt power from HV battery and to fuel pump).

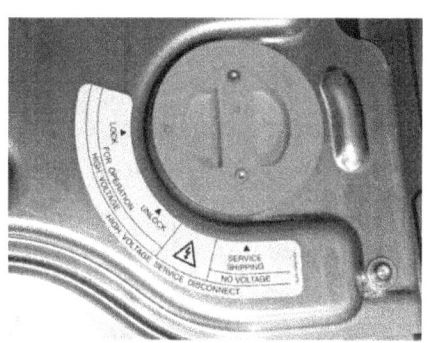

| High-Voltage Battery Pack | High-Voltage Service Disconnect Switch | High-Voltage Shut-Off Switch Located Behind Jack Access Panel |

Ford Fusion/Mercury Milan Hybrid: Service Plug location is behind rear seat on HV battery housing.

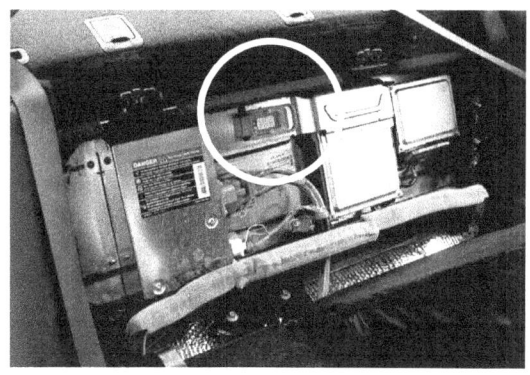

Honda Hybrids:

Honda hybrids typically <u>do not</u> have a removable Service Plug.

Instead of a removable Service Plug, Honda hybrids have a Battery Module Switch located behind rear seat on HV battery housing.

Remove cover shown to access the Battery Module Switch.

IPU Box Lid   Rear Seat-Back

Battery Module   Bolts
Switch Cover

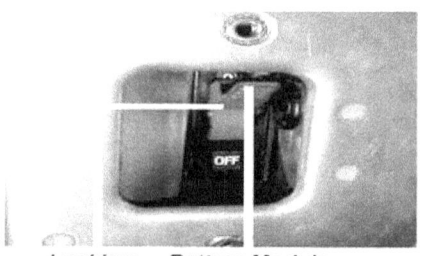

Locking   Battery Module
Cover   Switch

GM Malibu/Aura/Vue mild hybrids:

GM mild hybrids typically do not have a removable Service Plug.

Battery Disconnect Switch location is behind rear seat on HV battery housing. When the cover shown is unbolted and opened, the switch opens the HV battery circuit.

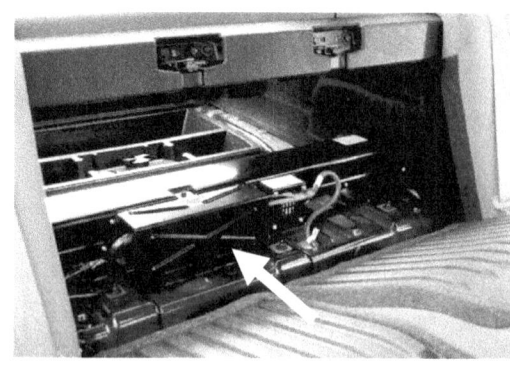

GM Tahoe/Yukon/Silverado/Sierra/ Escalade Two-mode hybrids: Service Plug location is under rear seat on left side.

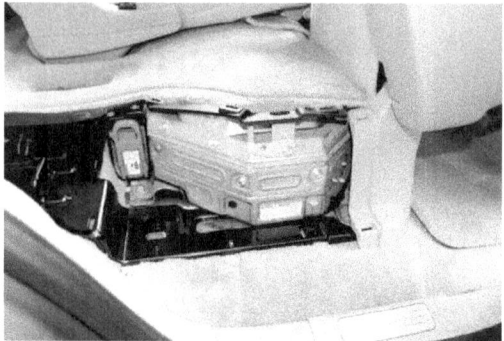

Service Plug removal and installation

Service Plug removal is a necessary safety procedure in order to disable the hybrid system and to remove high voltage from hybrid system components. The Service Plug must be removed before beginning work on any components in a high voltage system.

The Service Plug is typically located near the HV battery. This important procedure must be performed correctly to ensure that the HV battery is isolated from the hybrid system.

CAUTION!

Use extreme care when removing Service Plug. Incorrect Service Plug removal may create a shock hazard and may damage electronic control units.

Note: Disconnecting the 12-volt auxiliary battery may erase radio, GPS, telecom, infotainment, and other presets. Inform vehicle owner before proceeding.

Note: Removing the Service Plug may erase any Diagnostic Trouble Codes (DTCs) stored in ECU memory. Retrieve and record any DTCs from memory before removing Service Plug.

Typical Service Plug removal procedure – Toyota Prius:

1. Be sure the hybrid system (ignition switch) is in the OFF position.

2. If present, remove and store the smart key at least 15 feet from the vehicle. For additional safety, disable the Smart Key function.

3. Remove trim and/or flooring to gain access to Service Plug and 12-volt auxiliary battery.

4. Disconnect the auxiliary 12-volt battery. For added safety, cover exposed auxiliary battery terminal and cable end with insulating tape.

5. Wearing lineman's gloves, remove the Service Plug. Typical procedure: Lift handle straight upward to unlatch; rotate handle downward 90 degrees; pull handle to remove Service Plug.

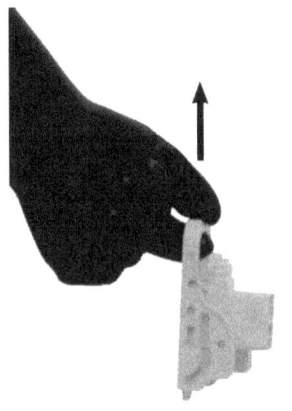

Step 1

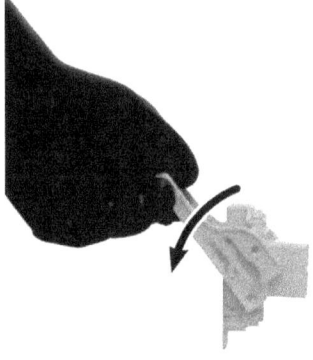

Step 2

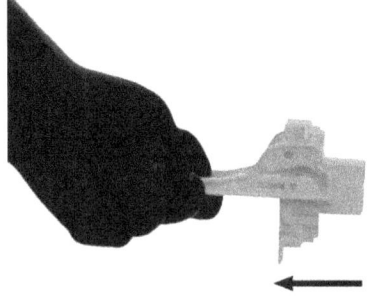

Step 3

6. To prevent unwanted Service Plug installation, store the Service Plug away from the vehicle.

7. For added safety, cover exposed Service Plug terminals on the HV battery with insulating tape.

8. Wait at least ten minutes for high-voltage system capacitors in inverter to dissipate voltage before proceeding. This will allow any charge in high voltage capacitors in the inverter to dissipate.

9. Test to ensure that high voltage is not present in any components that you may work on. Test procedures are described later in this Section.

Service Plug installation is the reverse of removal.

Service Plug removal - Ford Escape/Mercury Mariner Hybrid:

The Service Plug in Ford Escape/Mercury Mariner Hybrids is different from other designs.

To remove, wearing lineman's gloves, rotate orange plug counterclockwise and pull out.

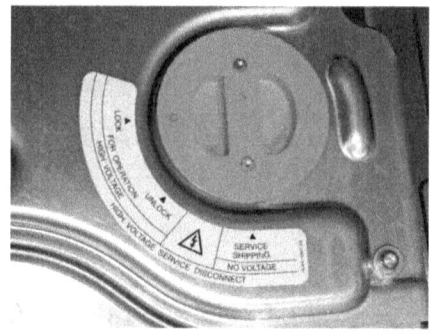

Service Plug removal - Honda hybrids:

There is no separately removable Service Plug on Honda hybrids.

Open cover shown on HV battery housing behind rear seat. Remove locking cover and throw Battery Module Switch to OFF.

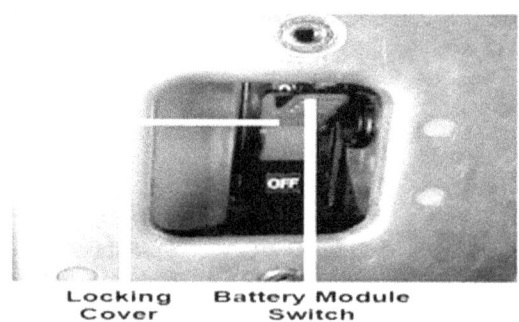

Service Plug removal - GM Malibu/Aura/Vue mild hybrids:

There is no separately removable Service Plug on the GM mild hybrids.

Opening cover shown on HV battery housing behind rear seat will release a plunger which will automatically open the integral HV Battery Disconnect Switch.

General procedure to make hybrid vehicles safe

SERVICE HINT:	Here is a general procedure to make hybrid vehicles safe to work on.

- Turn hybrid system OFF
- Disable Smart Key and store away from vehicle
- Disconnect 12-volt auxiliary battery
- Wear lineman's gloves
- Remove Service Plug
- Store Service Plug away from vehicle
- Test to ensure that high voltage is not present

Tests to ensure that high voltage is not present

 **CAUTION!**

Faults and abnormal conditions such as electrical shorts, grounds, and high resistance can cause potential high-voltage hazards. Remember, current flows because of a difference in voltage between two points. If you touch two different parts of the car that have very different voltage potentials, then your body becomes a conductor in a circuit and a shock may result!

Remember that in all vehicles (not just hybrids) the metal body/chassis of the vehicle and other metal components such as the engine provide the ground path for electric current for many circuits. Current can flow only when there is a complete path from the battery + (positive) terminal, through the circuit, and back to the battery – (negative) terminal. You may not think that the metal body acts like the insulated wires in a car, but it does. The body completes most circuits back to the battery. To prove this, just look at the battery negative cable, one end connects to the battery negative terminal and the other end to the vehicle body.

Note: In hybrid vehicles, the HV circuit is isolated from body ground and high voltage should never be present at the body ground. In case of accident damage in a Toyota hybrid for example, when there is significant HV harness, HV system, or body damage, the HV-ECU should open the SMR and shut down the HV system.

Electrical shorts can cause current to flow where it is not desired. Chafed or cut insulation can cause wire conductors to touch each other to create a short, or the short could be to ground.

High resistance can prevent high voltage from dissipating from components, creating a shock hazard. Also, there can be high-voltage capacitors in the inverter or other control units which may take up to ten minutes to dissipate their charge after power is removed.

Test to verify that high voltage is not present in any system that you want to work on.

All tests should be performed with good-quality equipment, such as a digital multimeter (DMM). The DMM and test leads should have a CAT III (1,000 volt) safety rating for testing hybrid high voltage circuits.

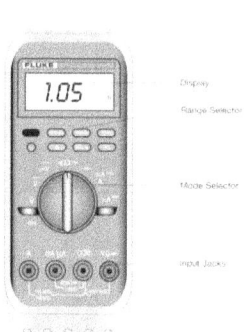

SERVICE HINT: 	Do not use an old style analog test meter (with a needle indicator). These do not have sufficient internal resistance and connecting them to electronic circuits may damage control modules.

CAUTION!

⚠️ ***Wear lineman's gloves until you have verified that high voltage is no longer present in components that you will be servicing.***

Note: Before performing any test to verify that high voltage is not present, always first disable the HV system by removing the Service Plug. The Service Plug removal procedure has been described earlier in this Section.

Voltage drop test:

A voltage drop test can help ensure that high voltage is not present in hybrid system components. To perform a voltage drop test, use a DMM. Touch test probes to any two suspected high voltage components (inverter, transaxle, and electric A/C compressor, for example) and measure voltage. If the meter does not auto-range, set for the highest expected voltage value and test. Reduce the range scale and test again until you have tested on the lowest scale.

Also measure voltage between high voltage components and a chassis ground. Be sure to touch the test probes to the case of each HV component being tested. All readings should be near zero volts.

Towing

Toyota 2WD hybrids: When towing a Prius or other hybrid with the THS/THS II type drive train (Nissan, Ford):

- Do not permit the drive wheels to rotate. The front wheels are directly geared to motor/generator MG2.

- MG2 turns any time the front wheels turn.

- Flat towing the vehicle can cause high voltage to be generated and can create a shock hazard.

- Flat towing may also create high voltage that may also damage the inverter and/or HV battery.

Toyota AWD hybrids: When towing AWD hybrids such as Toyota Highlander and Lexus RX450h:

- All four wheels should not be permitted to rotate. The front wheels are directly geared to motor/generator MG2 and the rear wheels are directly geared to MGR.

- MG2 and MGR turn any time the wheels turn.

SERVICE HINT: 	Prius vehicles with a discharged or disconnected 12-volt battery cannot be taken out of PARK position. The mechanism is electronically controlled and requires 12-volt auxiliary battery power.

Honda hybrids: Honda hybrids with the IMA hybrid system can be towed without danger of generating high voltage. As long as the engine is not running, and the transmission is in neutral, the engine crankshaft and IMA flywheel motor/generator components are stationary and will not generate voltage.

GM hybrids: GM mild hybrids, and can be towed without danger of generating high voltage.

Tow GM two-mode SUV and truck hybrids so that rear drive wheels are not permitted to rotate.

Disabling the HV system in an emergency

If you are unable to turn the hybrid system OFF with the ignition switch in the normal way, HV systems can be disabled by disconnecting or removing specified components. The procedure below applies to the 2004-2009 Prius. It is typical and similar to procedures for other hybrid vehicles.

 SERVICE HINT:	For more complete information about disabling the HV system in an emergency, see manufacturer's Emergency Response Guides in Appendix or online.

To disable the 2004-2009 Prius hybrid system:

- Disconnect the 12-volt auxiliary battery. Cover the battery terminal or cable connector with insulating tape to prevent connecting again accidentally.

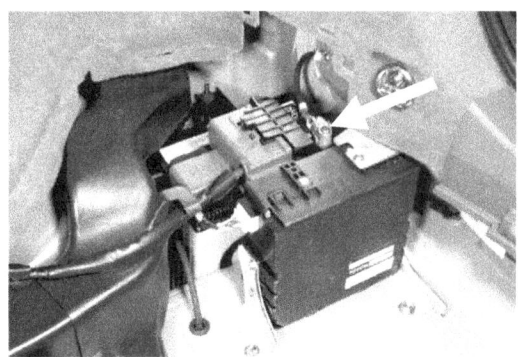

- Remove 20 amp yellow HEV fuse from engine compartment fuse/relay panel (junction block). If uncertain, remove all fuses from this panel.

(This page intentionally left blank)

Hybrid Powertrains

Learning objectives for this Section:

✓ You will be able to describe the four basic types of hybrid powertrains.
✓ You will be able to explain the overall operation of each hybrid powertrain type .
✓ You will be able to describe the operation of electrically powered auxiliary systems such as electric power steering and electric air conditioning compressors.

Introduction

This course discusses the largest selling hybrids. The hybrid powertrains in these vehicles can be grouped into four basic types:

- Toyota Hybrid System (THS) and Toyota Hybrid System II (THS II), also known as Toyota Hybrid Synergy Drive (similar systems are used by Lexus, Nissan and Ford)

- Honda Integrated Motor Assist (IMA)

- General Motors mild hybrid Belt Alternator Starter (BAS) system

- General Motors/BMW/Chrysler Two-Mode hybrid transmission

All four of these hybrid system types use electric traction motors to assist the gasoline internal combustion engine in order to improve mileage and reduce emissions. High voltage batteries store electrical energy and provide power to the electric motor.

The designs of each of these systems differ, but there are basic similarities and principles that apply to all. This section will describe how they are configured and how they work.

Toyota Hybrid System (THS) and Toyota Hybrid System II (THS II)

Engine

The 2001-2009 Prius uses the 1NZ-FXE 1.5 liter 4-cylinder dual overhead camshaft (DOHC) 16-valve engine with variable intake valve timing that has been modified to work with the hybrid system.

The 2010> Prius uses the larger 2ZR-FXE 1.8 liter version of this engine.

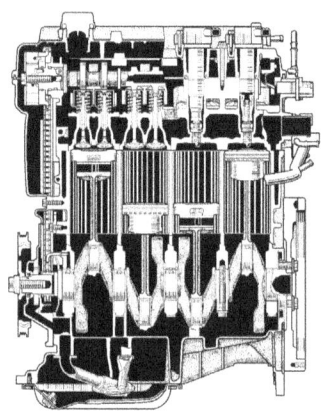 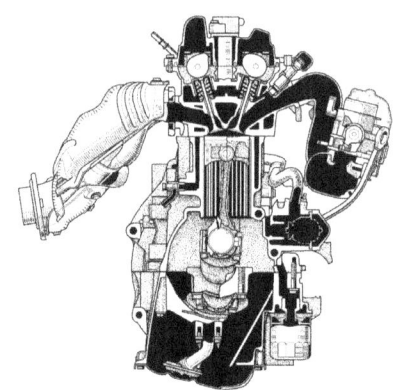

The engine is optimized for two roles:

* Provides power to the front wheels to drive the vehicle.

* Drives motor/generator MG1 as a generator to recharge the HV battery.

In the conventional 4-stroke cycle, the engine compression ratio (how much the intake mixture is compressed as the piston travels upward in the cylinder) and the expansion ratio (how much the spent gases expand during the power stroke as the piston travels downward) are the same.

The Prius engine operates on the Atkinson 4-stroke cycle. This means that the engine's compression ratio can be different from the expansion ratio.

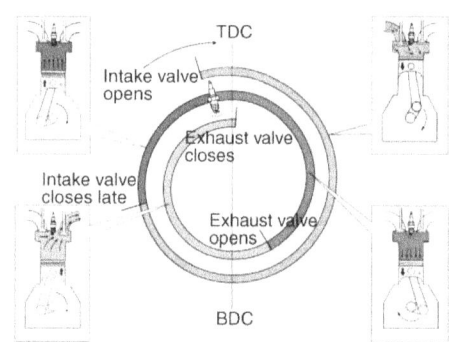

The variable valve timing system varies intake valve timing and can cause the intake valves to close very late during the intake stroke (and into the compression stroke). This effectively reduces the compression ratio but still allows full expansion during the power stroke. A benefit is increased engine torque. The calculated compression ratio is 13:1, but the late

intake valve closing effectively reduces this number.

Another effect of the Atkinson Cycle's late intake valve closing is reduced intake manifold vacuum. The rising piston pushes a small amount of the incoming intake charge (air and fuel) back out the intake valve and into the intake manifold, where it awaits the next cycle. The Prius engine has an enlarged intake manifold to accommodate this effect.

The Prius engine has its crankshaft offset 0.47 in. (12 mm) relative to the centerline of the cylinder bores. This reduces friction and side loads on pistons. The result is less engine wear and reduced noise.

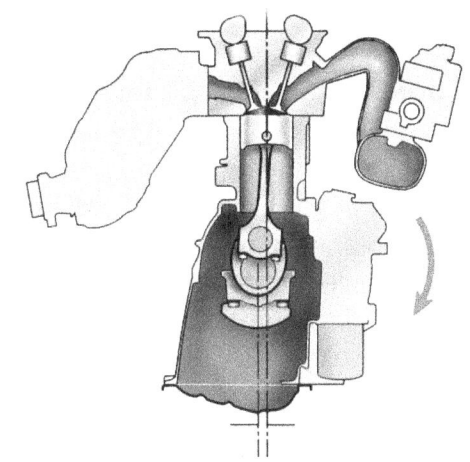

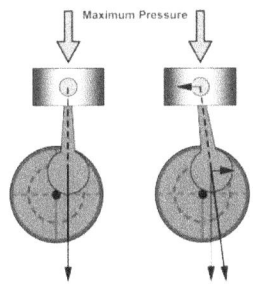

The Toyota Camry Hybrid uses the 2.5 liter 2AZ-FXE gasoline engine. Like the Prius, the Camry Hybrid engine is also an inline four cylinder, but larger.

The Toyota Highlander Hybrid uses the 3.3 liter 3MZ-FE V-6 gasoline engine.

Coolant heat storage tank

The engine cooling system in the 2004-2009 Prius has a unique container located in the left front fender that stores hot coolant for an extended period after the vehicle is turned OFF. When the engine is re-started, the hot coolant heats the intake manifold to promote fuel vaporization and to speed engine warm-up. This reduces hydrocarbon (HC) and carbon monoxide (CO) emissions.

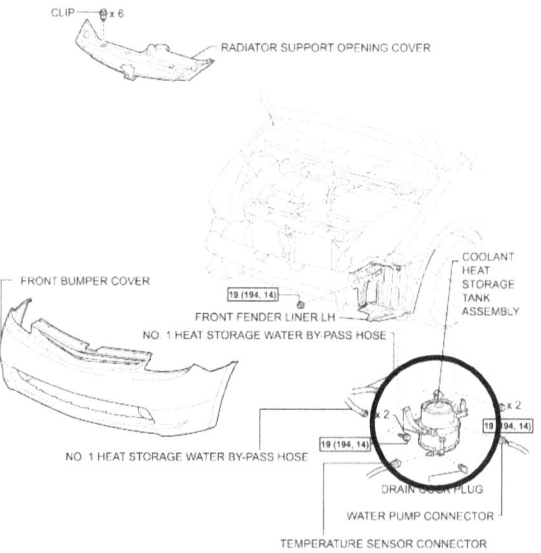

The coolant heat storage tank is made from sheet stainless steel and has double-wall construction. Vacuum between the tank walls insulates like a Thermos® bottle. The tank can store coolant as hot as 176°F (80°C) for up to three days.

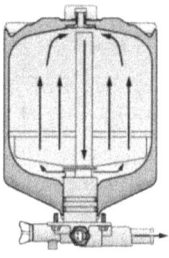

CAUTION!

The coolant heat storage tank in the 2004-2009 Prius stores coolant as hot as 176⁰F (80⁰C) for up to three days. Hot coolant can scald. Use extreme caution when draining coolant or when working near this tank.

THS/THS II transaxle

The Toyota Hybrid System transaxle has few moving parts compared with other manual and automatic transmissions. It contains two permanent magnet three-phase brushless high-voltage AC synchronous motor /generators (MG1 and MG2), and one planetary gear set.

MG1 and MG2 operate on 200-650 volts AC. Operating voltage varies with different THS versions and operating conditions.

The planetary gear set splits power to drive the wheels between the IC engine and motor/generator MG2. Other components include a silent chain, transfer gears, and the final drive differential.

No conventional starter motor is required. MG1 can be run as an electric motor to crank the IC engine for starting. And MG1 can be driven by the engine as a generator to provide additional current to power MG2, charge the HV battery, and charge the 12-volt auxiliary battery.

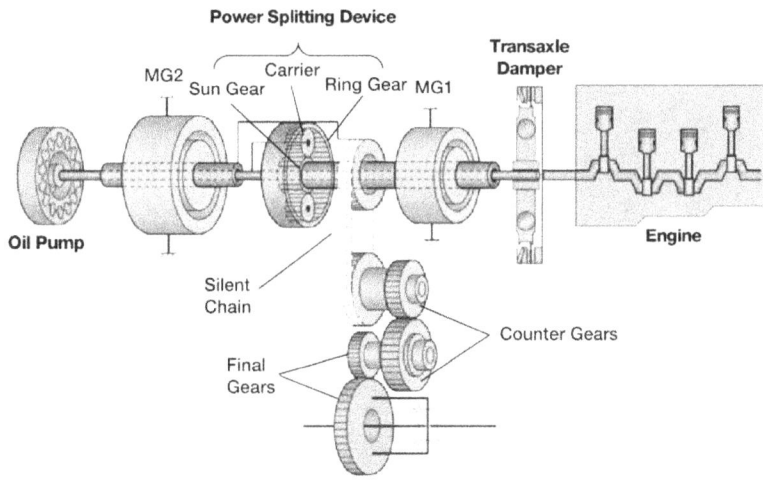

Toyota hybrid transaxle designations are as follows:

2001-2003 Prius:	P111
2004-2009 Prius:	P112
Highlander Hybrid:	P310
Camry Hybrid:	P311
2010> Prius:	P410

Toyota Highlander Hybrid models with AWD have a third motor/generator mounted in the rear and driving the rear wheels. This motor/generator is called MGR.

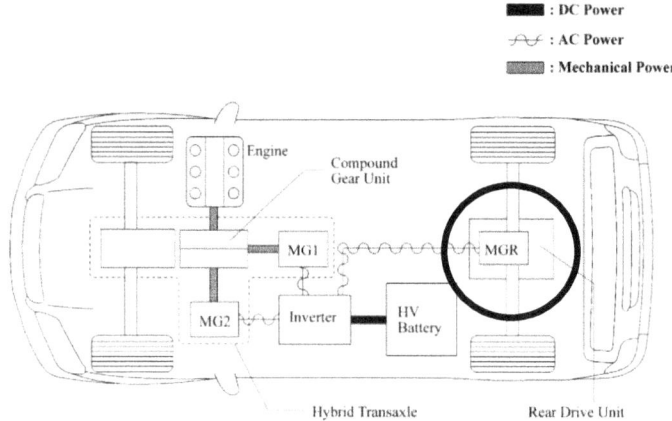

Hybrid Vehicle Electronic Control Unit (HV-ECU)

The HV-ECU manages and controls all hybrid system functions, including hybrid transaxle and engine operation. The HV-ECU also communicates with other control units over the CAN network.

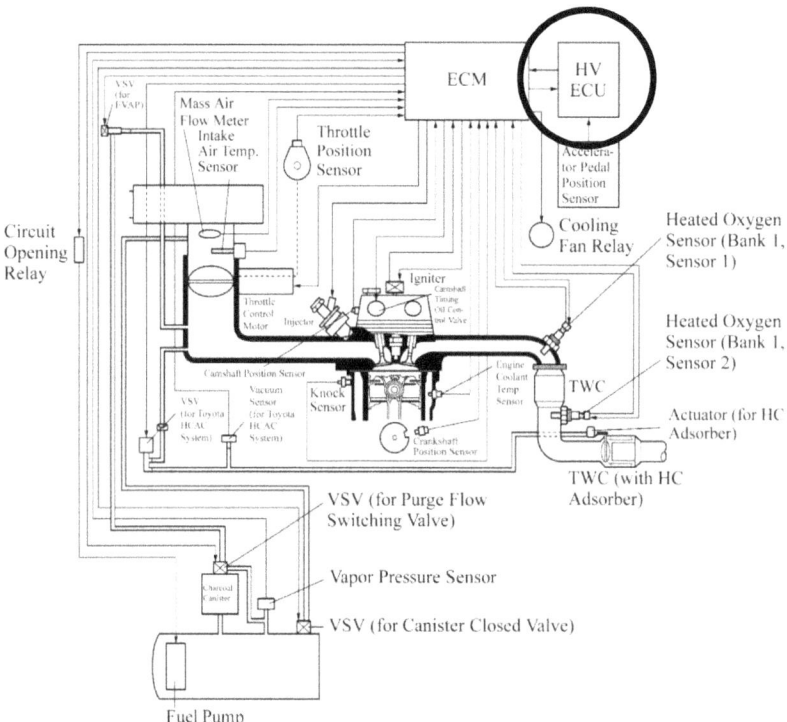

Inverter

The HV battery is a direct current (DC) device. Electric motor/generators MG1 and MG2 are three-phase alternating current (AC) devices. The inverter changes AC to DC and DC to AC.

The inverter changes DC voltage from the HV battery to three-phase AC to run MG2 as an electric motor.

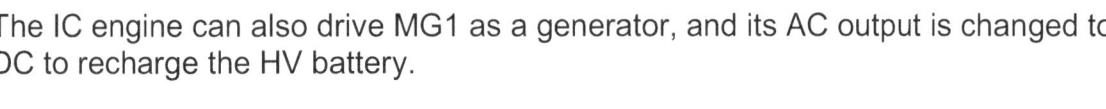

During regenerative braking, MG2 is driven as a generator by the wheels. The inverter changes the generator's three-phase AC output voltage to DC to recharge the HV battery.

The IC engine can also drive MG1 as a generator, and its AC output is changed to DC to recharge the HV battery.

Boost Converter

The boost converter in the THS II inverter can increase system voltage from the HV battery to MG2 up to 500 volts (Prius) and up to 650 volts (Camry, Altima, and Highlander). This has allowed smaller HV batteries when compared with earlier hybrid models.

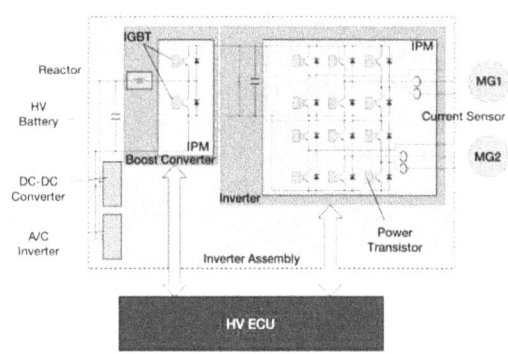

DC-DC Converter

When the hybrid system is ON, a DC-DC converter in the inverter transforms high DC voltage from the HV battery to 12 volts DC to operate 12-volt systems and accessories. This 12-volt power recharges the 12-volt auxiliary battery, powers lights, audio system, A/C blower fan, and other 12-volt accessories.

Nissan Hybrid System

The Nissan Altima Hybrid uses the Toyota Camry P311 THS transaxle mated to the Nissan 2.5 liter 4-cyl QR25DE engine.

The Altima Hybrid also uses the Toyota THS HV battery, inverter and electronic controls.

The 12-volt auxiliary battery is a conventional lead-acid battery. This battery only provides voltage to power-up control modules when turning the vehicle ON.

Ford Hybrid System

Ford Escape and Fusion Hybrid models (along with corresponding Mercury Mariner and Milan versions) have Ford's own version of the Toyota THS transaxle mated to a 2.5 liter Atkinson cycle I-4 engine.

The Ford hybrid transaxle design similar to Toyota's THS hybrid transaxle, but it is unique to Ford and built by Aisin.

Honda Integrated Motor Assist (IMA) Hybrid System

All Honda hybrid vehicles use the Honda Integrated Motor Assist (IMA) system.

IMA is a brushless AC electric motor/generator in place of the engine flywheel. The IMA motor/generator is the starter for the IC engine. It provides electric drive assist to the wheels, and acts as a generator during regenerative braking.

In the Civic model, the IMA motor is rated at 13 hp and 46 ft/lbs torque.

The hybrid control unit in the Honda is called the Intelligent Power Unit (IPU).

The first hybrid vehicle sold in the USA was the 1st generation Honda Insight in 2000.

The engine in the 2000-2006 Insight is a 1 liter I-3 engine mated to either a 5-speed manual or automatic CVT (Continuously Variable Transmission).

Other Honda hybrids use either I-4 (Civic Hybrid) or V-6 (Accord Hybrid) engines.

GM Mild Hybrid System – Belt Alternator Starter (BAS)

Chevrolet Malibu, Saturn Aura, and Saturn Vue Hybrids have a mild hybrid Belt Alternator Starter (BAS) system. The BAS system replaces the conventional alternator (generator) with a combined starter motor/generator unit. The electric motor is rated at 3 kW.

Main benefits include engine-off-at-stop and during deceleration. Regenerative braking also recaptures some energy to recharge the 36-volt HV battery. There is mild assist during acceleration from the electric motor.

The BAS motor/generator unit is driven by a multi-rib belt. A special tensioner keeps the belt tight. The engine in these vehicles is the 2.4 liter Ecotec I-4.

The transaxle is the 4T45-E that has been modified with a different final drive ratio. Since the engine may turn OFF when the vehicle stops, the transaxle also has an electric oil pump to maintain constant transmission oil pressure while the engine is OFF.

These vehicles are no longer in production.

GM Two-mode Hybrid System

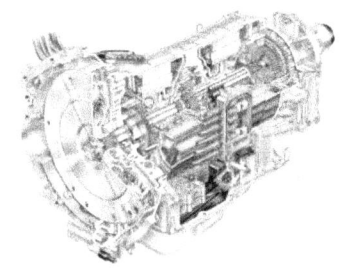

Chevrolet Silverado and Tahoe, GMC Yukon and Sierra, and Cadillac Escalade Hybrid SUVs and trucks mate a two-mode hybrid automatic transmission with a 6-liter V-8 engine. The transmission was co-developed with BMW and Chrysler.

The vehicle can be driven by the engine, by the electric motor, or both. The HV battery is rated at 300 volts. Regenerative braking is possible.

The transmission has two modes of operation:

Mode 1: Low speed, light loads. The vehicle can move under electric power alone, or both electric and IC engine power.

Mode 2: Higher speed and load. Electric power supplements IC engine power.

An electric oil pump in the transmission maintains constant transmission oil pressure while the engine is OFF.

Electric auxiliary systems

Electric power steering

Electric power steering (EPS) is typically used in hybrid vehicles for several reasons. First, electric power steering is more efficient and requires less power than conventional engine-driven hydraulic power steering. Second, engine-driven hydraulic power steering will not provide assist under conditions when the engine is OFF, such as during deceleration or when stopped.

GM two-mode hybrids have a 42-volt electro-hydraulic power steering pump which can run independently from the engine.

Electric power steering systems operate with voltages between 12-42 volts, depending on manufacturer and model:

12 volts:	Toyota Prius; Honda Hybrids; Ford Hybrids
36 volts:	GM mild hybrids - Malibu/Aura/Vue
42 volts:	Toyota Camry and Highlander Hybrids; Nissan Altima Hybrid; GM Two-mode Hybrids

The steering electric assist motor may be integral with the steering rack, mounted low on the steering column, or high on the steering column behind the instrument panel.

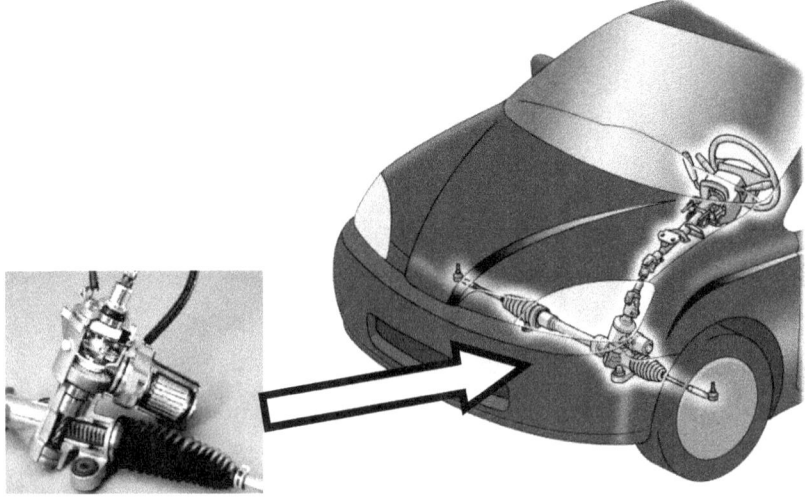

2000-2003 Prius

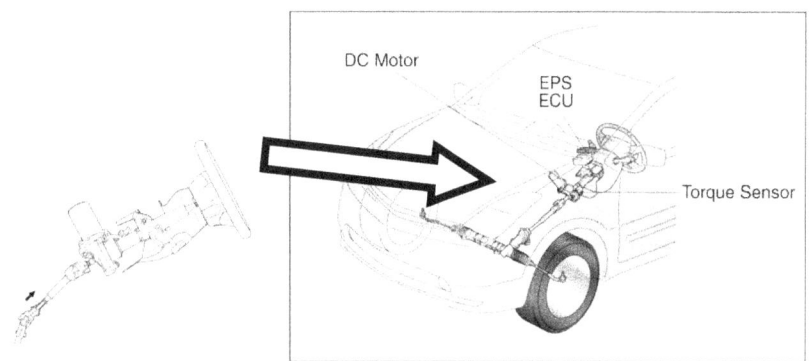

2004-2009 Prius

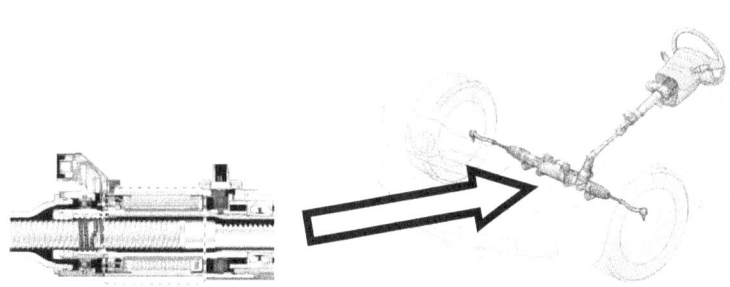

Toyota Camry, Nissan Altima Hybrids

Electric air conditioning compressor

The air conditioning (A/C) compressor is typically belt-driven by the engine. Many hybrid vehicles have electric A/C compressors that can run continuously independent of engine operation.

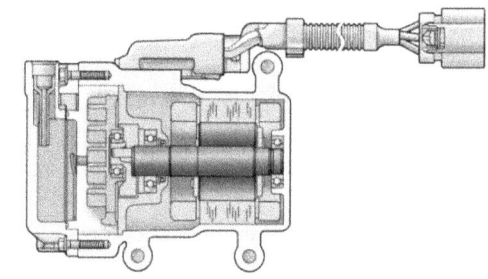

Toyota uses scroll-type compressors with a brushless high-voltage 'wet' electric motor. The motor is wet because refrigerant and refrigerant oil bathe and flow around the motor armature.

There are several important concerns about these electric compressors:

- Electric A/C compressor and wiring are a shock hazard. Compressor voltage:

    o 2004-2009 Prius: 201.6 volts

    o Camry Hybrid 244.8 volts

    o GM two-mode hybrid: 300 volts

- The high-voltage A/C compressor motor requires using a special high-dielectric refrigerant oil. It is important to use only the specified refrigerant oil. Toyota specifies oil type ND Oil 11. If another refrigerant oil is introduced to the A/C system, reduced insulation properties can reduce performance and create a potential shock condition.

CAUTION!

 **Electric A/C compressors are 200-300 volt high-voltage systems. Be sure hybrid system is OFF and high voltage is not present before starting work.**

The Honda Accord Hybrid has a dual-scroll compressor that is driven by both the engine with a belt, and also an internal electric motor.

Ford Fusion/Mercury Milan hybrids are equipped with an electric A/C compressor.

GM mild hybrids - Malibu/Aura/Vue have a 36 volt electric A/C compressor.

Maintenance and Repairs

Learning objectives for this Section:

✓ You will be able to list common maintenance and service related hybrid vehicle hints and tips.
✓ You will be able to perform common hybrid vehicle maintenance and service procedures after correctly ensuring that the HV system is safe.

Introduction

This Section contains a selection of common hybrid vehicle maintenance and repair procedures. This information is based on real-world knowledge and feedback about the most commonly occurring failures and required maintenance.

CAUTION!

When the hybrid system is ON and in the Ready Mode or Auto Stop Mode, the IC engine may start at any time. Be sure to turn the hybrid system OFF before performing any service.

Prius - Inspection Mode

Inspection Mode: Inspection Mode is an operating mode that allows the Prius IC engine to run continuously even if the vehicle is not moving. Official state inspection procedures may require the engine to run and this can be accomplished with Inspection Mode.

During normal operation, the IC engine is controlled by the Hybrid Vehicle Electronic Control Unit (HV ECU), and the engine may or may not run when in Ready Mode and the vehicle is not in motion. Inspection Mode commands the engine to run continuously.

Inspection Mode can be enabled with or without the Toyota scan tool (Techstream). The following describes how to enter Inspection Mode without a scan tool:

1. Engine should be warm, set parking brake, gear selector in P, and A/C OFF.

2. Turn ignition ON (READY lamp illuminated), then turn ignition OFF.

3. Perform the following steps within 60 seconds.

4. DO NOT depress the brake pedal. Press and release the Power button twice to enter IG-ON.

5. Fully depress and release the accelerator pedal 2x with the gear selector in P position.

6. Depress the brake pedal and move the gear selector to N position.

7. Fully depress and release the accelerator pedal 2x with the gear selector in N position.

8. Depress the brake pedal and move the gear selector back to P position.

9. Fully depress and release the accelerator pedal 2x with the gear selector in P position.

10. The hybrid system warning on the multi-function display should flash.

11. Depress the brake pedal and depress and release the Power button once to turn ignition ON (READY lamp illuminated). The IC engine should start and run continuously.

When in Inspection Mode, engine speed depends on accelerator pedal position:

- Accelerator pedal not depressed: idle (approx. 1000 rpm)

- Accelerator pedal depressed <60%:   approx. 1500 rpm

- Accelerator pedal depressed >60%:   approx. 2250 rpm

Alternative to Inspection Mode: There is another way to force the Prius IC engine to run continuously. Although this method does not permit the engine to idle, it can be used to quickly command the engine to run:

1. Set parking brake, gear selector in P, and A/C OFF.

2. Turn ignition ON (READY lamp illuminated).

3. With gear selector in P, simply depress the accelerator pedal. The engine will run continuously as long as the pedal is depressed. Releasing the pedal usually causes the engine to stop.

Prius - Engine

Oil specification: The stop-start duty cycle of the Prius engine, plus the use of engine oil as a hydraulic fluid in the variable valve timing (VVT-i) system, make both oil specification and oil quantity critical.

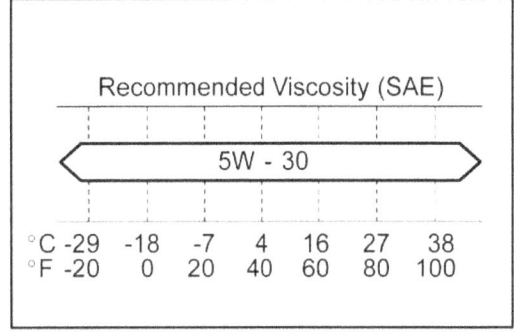

Recommended Viscosity (SAE)

5W - 30

| °C | -29 | -18 | -7 | 4 | 16 | 27 | 38 |
| °F | -20 | 0 | 20 | 40 | 60 | 80 | 100 |

For 2004-2009 Prius models, use only good quality SAE 5W-30 oil and do not overfill.

Using incorrect specification oil may cause a DTC to set.

SERVICE HINT: 	Overfilling engine oil may cause oil to collect in the intake manifold. This may cause a no-start condition and HV battery discharge.

Compression testing: Since the Prius engine does not have a conventional cranking (starter) motor, the procedure for performing a compression test is different.

To perform a compression test, MG1 can be commanded to crank the engine with a Toyota scan tool or equivalent. The Toyota scan tool has an ACTIVE TEST that will cause MG1 to crank the engine over continuously. The basic procedure steps are:

- Warm engine
- Gain access to top of engine
- Disable fuel injectors
- Remove all ignition coils and spark plugs
- Connect scan tool and select ACTIVE TEST > COMPRESSSION TEST
- Turn ignition ON
- Engine cranks continuously, measure compression as quickly as possible to avoid depleting HV battery

SERVICE HINT:	The HV battery cannot just "go dead." If the HV battery state-of-charge is very low (battery depleted), there may be a no-start condition.

Engine access: Some engine compartment repair procedures require removal of the cowl and windshield wiper motor assembly for access. Example: engine valve adjustment.

Ignition coils and spark plugs can be removed without removing cowl.

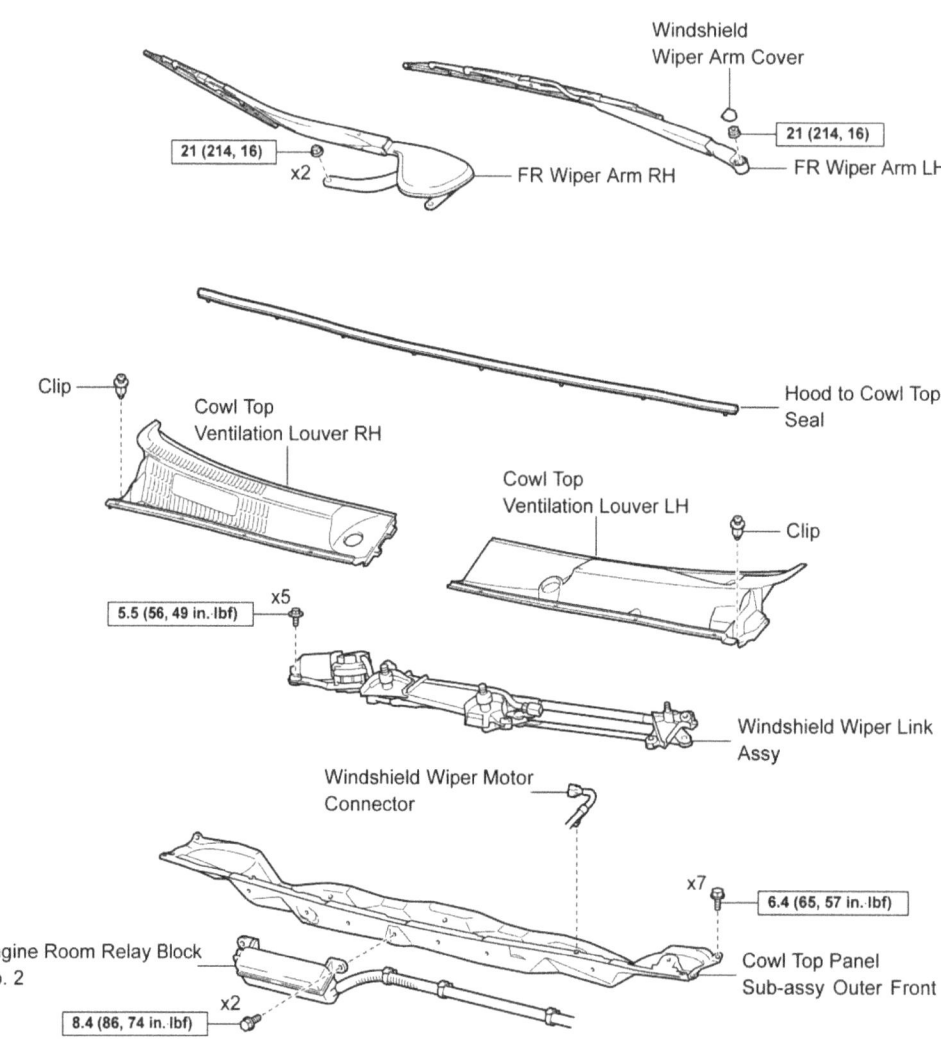

Windshield Wiper Arm Cover

21 (214, 16)

21 (214, 16)
x2

FR Wiper Arm RH

FR Wiper Arm LH

Clip

Cowl Top Ventilation Louver RH

Hood to Cowl Top Seal

Cowl Top Ventilation Louver LH

Clip

5.5 (56, 49 in. lbf) x5

Windshield Wiper Link Assy

Windshield Wiper Motor Connector

x7

6.4 (65, 57 in. lbf)

Engine Room Relay Block No. 2

Cowl Top Panel Sub-assy Outer Front

x2

8.4 (86, 74 in. lbf)

Engine cooling system: On 2004-2009 Prius models, the serpentine ribbed accessory drive belt drives only one component, the engine coolant pump. The belt is therefore a narrow light-duty belt.

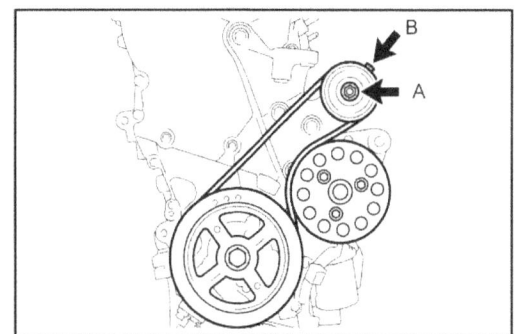

Engine coolant pump: Replacing the Prius engine coolant pump is a common repair.

Leaking coolant pumps can be diagnosed by a pink coolant "stripe" under the hood, a wet coolant pump pulley, and loss of coolant from the reservoir.

Note: A slight amount of coolant loss is considered normal.

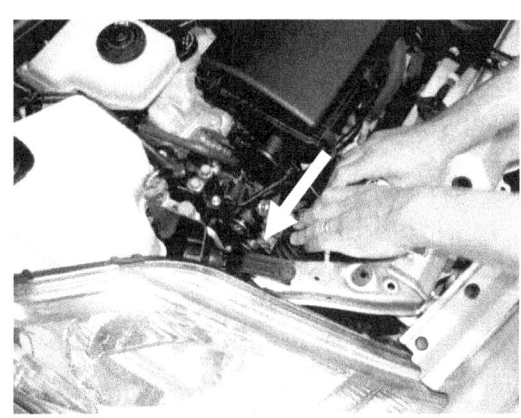

Coolant heat storage tank electric pump: 2004-2009 Prius models have the coolant heat storage tank in the right front fender. The tank has an electric pump that can become noisy.

There is a Technical Service Bulletin (TSB) from Toyota for replacement of a noisy pump. The TSB number is: T-SB-0087-08.

The Prius engine cooling system is more complex than most vehicles. In addition to the engine-driven coolant pump, there are two electric coolant pumps and an electrically operated rotary coolant flow control valve.

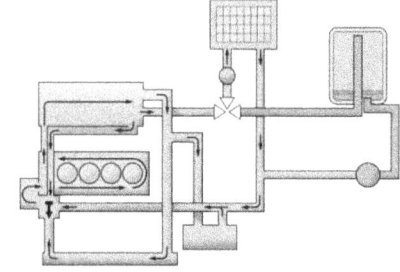

Engine coolant drains: Engine cooling system drains for the 2004-2009 Prius are shown in this graphic.

Toyota service information (TIS) specifies that the coolant heat storage tank pump connector be disconnected when draining engine coolant. This is done to prevent the electric pump from running when the system is empty. Disconnecting this pump may cause a DTC to set and the Malfunction Indicator Lamp (MIL) to illuminate.

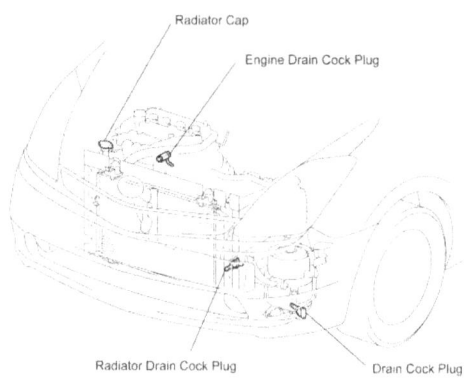

If a DTC is set and the MIL illuminates after engine coolant drain and refill, use a Techstream scan tool or equivalent to clear the DTC.

Refilling engine cooling system: Refilling and bleeding air from the Prius engine cooling system after draining can be challenging. Toyota service information (TIS) specifies a procedure that uses the Techstream scan tool or equivalent to run the electric coolant pump. The engine must experience several heating and cooling cycles while the system is bled. This can be time consuming.

An alternative method for refilling the engine cooling system uses a cooling system vacuum fill tool. One brand on the market is called AirLift. When powered by a shop air supply, the tool creates a vacuum that will remove most air from the cooling system. Large diameter radiator hoses may collapse during this process.

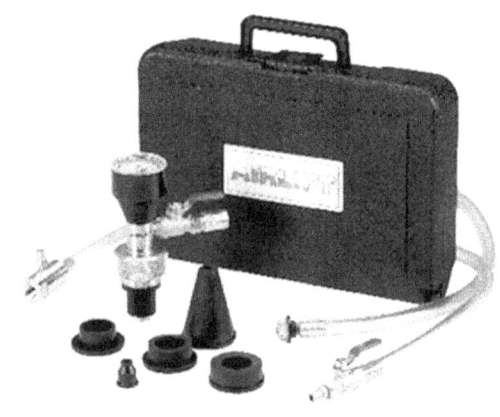

Closing one valve on the tool and opening a second valve allows atmospheric pressure to push coolant from a supply bottle into the cooling system. A system refilled this way often fills completely and requires no bleeding.

Vacuum fill tool attaches to radiator filler neck. Be sure to refill with the specified coolant.

2010> Prius electric coolant pump:

The 2010> Prius does not use an engine-driven coolant pump. The 2ZR-FXE engine in this model has an electric coolant pump. The pump uses a brushless-type DC motor controlled by the ECM. On the Toyota Techstream scan tool, this pump is displayed as Coolant Pump B.

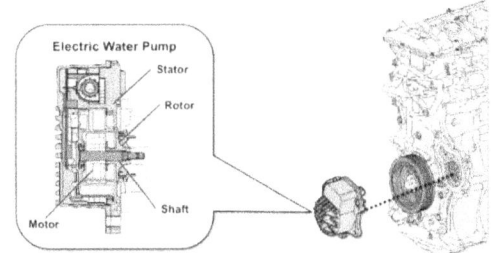

There is no crankshaft-driven belt on this engine since all of the accessories that are normally belt-driven are electric (power steering, A/C, coolant pump, alternator, etc.).

Toyota – THS/THS II transaxle cooling system

The Toyota THS and THS II transaxles have a liquid cooling system that is similar to and completely separate from the engine cooling system.

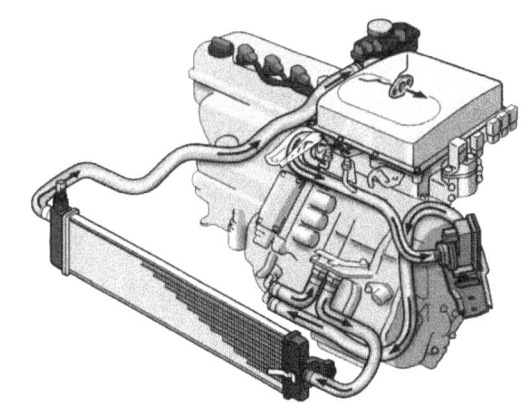

The transaxle cooling system has a dedicated radiator, reservoir, electric pump, and lines. Coolant from this system does not mix with engine coolant, although the same type of coolant is used.

The transaxle cooling system also cools the inverter, which mounts on top of the transaxle and has coolant passages in its base.

Draining and refilling transaxle cooling system: When draining transaxle coolant, be sure to open the correct drain plug. See graphic below. Replace drain plug seal ring.

Coolant Drain · Trans Fluid Fill · Trans Fluid Drain

Transaxle cooling systems on all Prius models can be difficult to refill with coolant and bleed. Toyota service information (TIS) specifies a procedure that allows the transaxle cooling system electric pump to circulate coolant while a bleeder is opened. This procedure is repeated several times and can be time consuming.

An alternative method for refilling the transaxle cooling system uses a cooling system vacuum fill tool, such as the AirLift brand tool. The basic procedure is the same as described under **Refilling engine cooling system** in this Section.

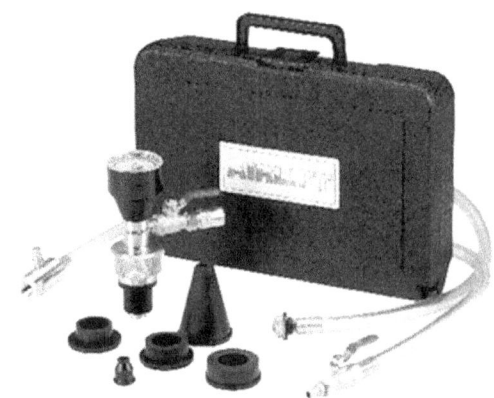

Vacuum fill tool attaches to transaxle coolant reservoir filler neck. Be sure to refill with the specified coolant.

Toyota – THS/THS II transaxle fluid service

The Prius maintenance schedule does not suggest a fluid change interval for transaxle lubricant.

Industry feedback suggests that draining and refilling THS/THS II transaxle fluid periodically is a good maintenance practice.

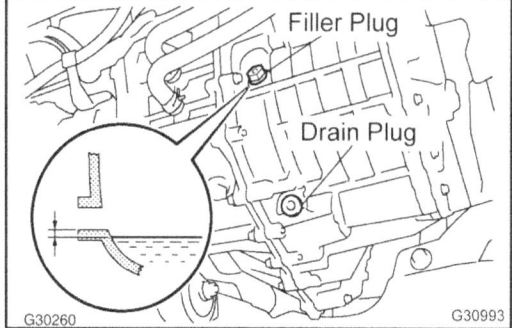

SERVICE HINT:	Be sure to identify correct drain and fill plugs. Be sure to use the factory-recommended transmission fluid.

To refill transaxle fluid, a small diameter plastic hose with a funnel can be easily inserted into the open transaxle fill hole.

Use the specified fluid and replace seal rings. Do not overfill.

High Voltage (HV) batteries

Nickel metal hydride (NiMH) HV batteries have proven to be reliable and do not require frequent service or replacement. The hybrid system controls HV battery charge carefully and never allows the HV battery to become fully charged or discharged, which helps lengthen HV battery life.

Some early Prius models (2001-2003) experienced corroded HV battery terminals where the bus bars are bolted to the battery cell terminals.

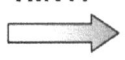**SERVICE HINT:**	Toyota has a reward program for recycling HV batteries and pays a "bounty" to dealers for returned batteries. See your local dealer if HV battery disposal is necessary.

SERVICE HINT:	The HV battery cannot just "go dead." If the HV battery state-of-charge is very low (battery depleted), there may be a no-start condition.

Prius - 12 volt auxiliary battery

Most hybrids have a conventional 12-volt battery in addition to the high voltage hybrid system battery. The 12-volt auxiliary battery in the Prius is a special absorbed glass mat (AGM) type lead-acid maintenance free battery with low current capacity.

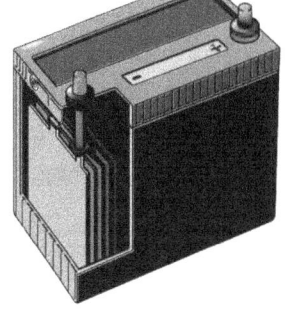

The Prius auxiliary battery cannot provide high current like a conventional flooded-cell lead-acid battery, but it does not need to provide high current. The 12-volt auxiliary battery powers electronic control units (ECUs) and 12-volt systems during start-up only. This battery does not provide high current and does not power a cranking (starter) motor.

Since the Prius battery is in a location that is open to the passenger compartment, the battery is vented to outside air through a plastic hose. Be sure the hose is properly installed and routed.

There have been auxiliary battery failures reported in the Prius. The vehicle will not start and will not enter Ready Mode without a good 12-volt auxiliary battery.

If charging is required, charge with low current only. Do not exceed a 3.5 amp maximum charging rate.

SERVICE HINT: 	12-volt auxiliary battery problems can cause DTCs, improper or erratic operation of vehicle systems, and no-start conditions. Always begin diagnosis by checking auxiliary battery condition.

65

Prius - Jump starting

The 12-volt auxiliary battery in a Prius is located in the rear cargo area and can be jumped remotely.

There is a positive jumper terminal on the junction/relay block, located on the left side of the engine compartment.

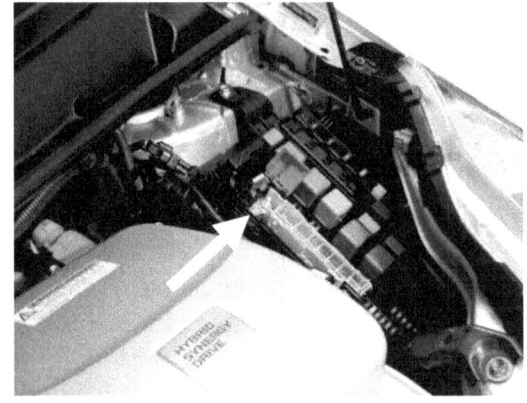

A convenient ground point is located on the right side of the engine compartment.

Follow all safety precautions when jump starting a Prius.

SERVICE HINT: 	Do not use the positive jumper terminal on the junction/relay block to recharge a weak 12-volt auxiliary battery.

Prius – Air/Fuel Ratio (AFR) sensors

In the 2004-2009 Prius, the Air/Fuel Ratio (AFR) sensor in the B1S1 location (ahead of the catalytic converter) has been known to set fault codes (DTCs).

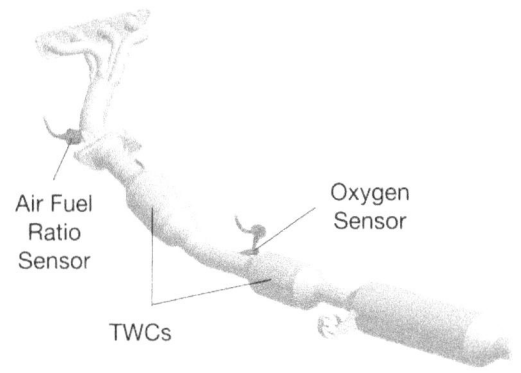

The AFR sensor connector is under the inverter (Arrow). Inverter removal is a long procedure which requires opening the inverter, disconnecting high voltage circuits and draining the transaxle/inverter cooling system. Consult factory service information before attempting this repair.

 **SERVICE HINT:**	Toyota service information says that AFR sensor removal requires inverter removal in order to access the sensor connector. Toyota TSB PG002-06 documents inverter removal cautions.

Prius - Oxygen (O2S) sensors

The oxygen sensor (O2S) in the B1S2 location is the sensor located after the catalytic converter.

On a 2004-2009 Prius, the sensor connector is located in the passenger compartment on the right side of center console under the carpet.

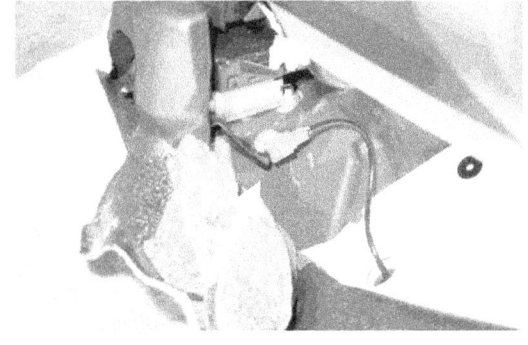

Prius - Xenon High Intensity Discharge (HID) Headlamps

Both halogen and Xenon HID headlamps have been available on Prius models.

The HID headlamps on 2004-2009 Prius models reportedly have durability problems.

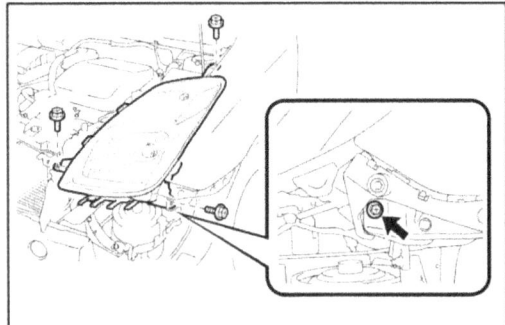

Toyota service information says that, on 2004-2009 Prius models, the bumper cover and headlamp assembly must be removed to replace headlamp bulbs.

SERVICE HINT: 	The HID headlamps are very sensitive to bulb quality. Use only factory –type replacement bulbs. HID bulbs should be replaced in pairs to ensure proper system operation.

Industry feedback suggests that both halogen and HID headlamp bulbs can be replaced with the headlamp assembly in the car. Space is limited, but it can be done

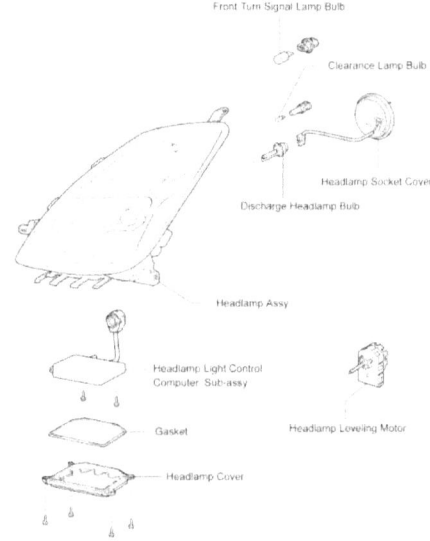

Front Turn Signal Lamp Bulb
Clearance Lamp Bulb
Headlamp Socket Cover
Discharge Headlamp Bulb
Headlamp Assy
Headlamp Light Control Computer Sub-assy
Headlamp Leveling Motor
Gasket
Headlamp Cover

CAUTION!

HID headlamp circuits can produce up to 20,000 volts for short periods. Shock hazard! Be sure hybrid system is OFF before performing any headlamp service.

Prius - Exhaust system

The 2001-2003 Prius exhaust system has an unusual catalytic converter. On this model, the normal catalyst element is surrounded by a device called the HC Adsorption Catalyst (HCAC) system. This device is designed to trap and hold hydrocarbon emissions 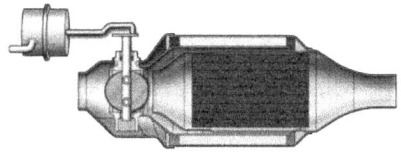 (HC) during cold starts, before the catalyst has "lit off" and before it has begun to work. This is one way to reduce HC emissions when they are especially high.

Once the catalyst has become hot enough for the oxidation and reduction reactions to take place, the adsorber purges the stored HC.

The HCAC system has moving parts. Exhaust flow through the adsorber element is controlled by a vacuum operated flow control valve mechanism.

 **SERVICE HINT:**	The HCAC flow control valve can stick and can require maintenance or repair. This mechanism can be affected by wear, corrosive road salt and thermal cycles.

The 2010> Prius also has an unconventional exhaust system. An exhaust heat recirculation system has been added to speed coolant and engine warm up. Lines and hoses direct engine coolant to a coolant jacket built around the sub-muffler at the forward end of the exhaust pipe, just after the three-way-catalyst (TWC).

An exhaust pipe gas control actuator controls the flow control valve. When engine coolant temperature is below approx. 160°F, the exhaust gas control valve is closed. Exhaust gases are forced to flow around a coolant jacket before continuing down the exhaust pipe.

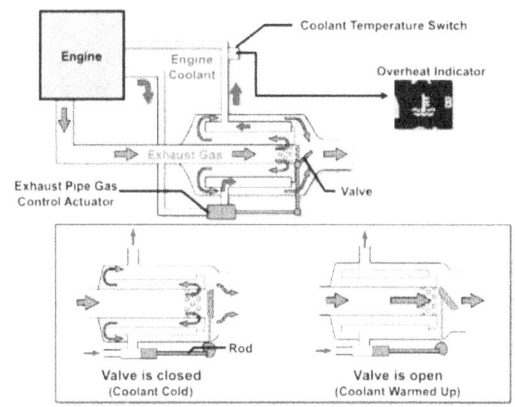
Exhaust Heat Recirculation System

Once engine coolant temperature exceeds 160-170°F, the actuator opens the exhaust gas control valve and exhaust gases flow straight through the exhaust pipe.

A dedicated coolant temperature switch monitors coolant temperature as it exits the exhaust pipe assembly. If the exhaust gas control valve fails in the closed position and overheats the engine coolant, a warning indicator will be displayed on the combination meter.

While the 2010 Prius is a new model vehicle, the undercar moving parts for this system are likely to be adversely affected by road conditions and require maintenance or repair.

Prius - Brakes

The regenerative braking system on the Prius uses conventional disk and drum brake assemblies at the wheels. These brake assemblies can be maintained and serviced like any other conventional braking system. Brake pad and shoe replacement intervals tend to be long since regenerative braking contributes a significant amount of the braking force.

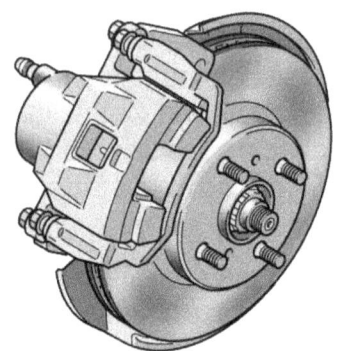

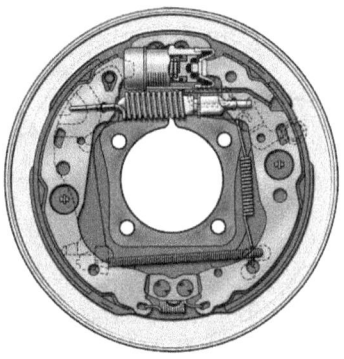

When the brake pedal is depressed in the 2004-2009 Prius, the Skid Control ECU controls the braking contribution from regenerative braking, and also from hydraulic braking.

Braking may feel like conventional hydraulic braking, but regenerative braking (allowing the wheels to turn MG2 as a generator) contributes much of the force to slow the vehicle.

Brake power backup: 2004> Prius vehicles have a brake power source backup unit located in the right rear quarter next to the 12-volt auxiliary battery.

The brake power source backup unit contains 28 capacitors that store a 12-volt charge. This provides a power supply for brake system if the normal vehicle 12-volt power supply is interrupted.

When the brake pedal is depressed during normal braking, fluid sent to each wheel does not come from the brake master cylinder. Rather, fluid quantity and pressure to each wheel is controlled by the Skid Control ECU.

This diagram shows brake fluid flow during normal braking:

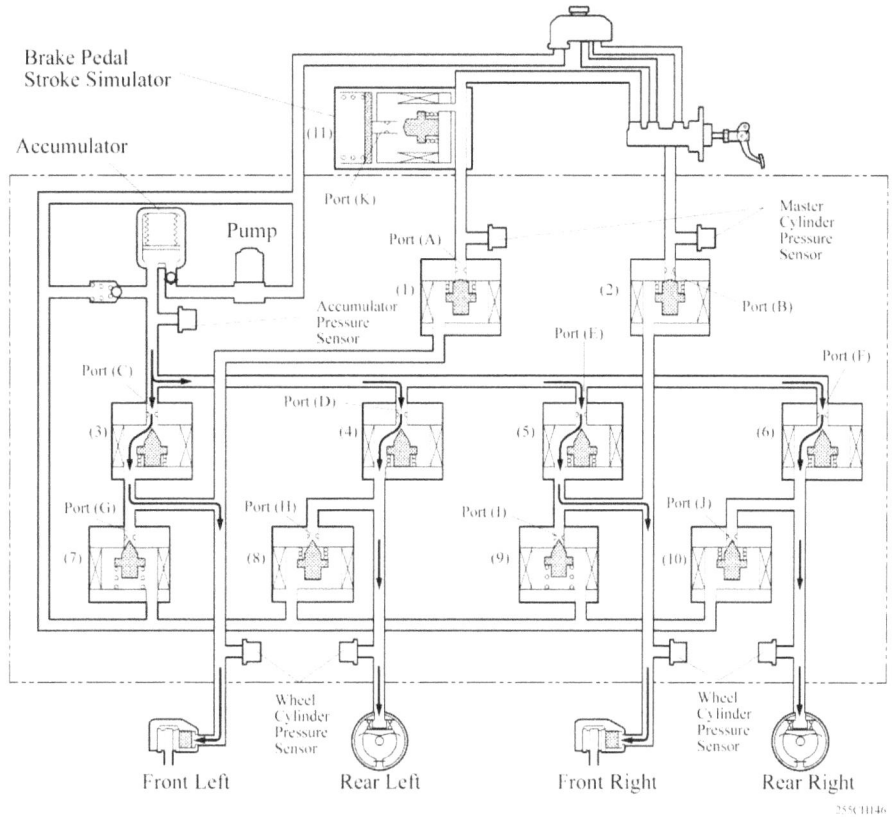

SERVICE HINT:	Prius brakes must be bled by using a Toyota Techstream scan tool or equivalent.

SERVICE HINT:	Procedures such as installing new brake pads and retracting caliper pistons may cause DTCs to set and the MIL to illuminate after brake pad replacement. If a DTC is set and the MIL illuminates after brake pad replacement, use a Techstream scan tool or equivalent to clear the DTC. Be sure the hybrid system is OFF during service.

71

Toyota - Diagnostic information and Diagnostic Trouble Codes (DTCs)

The On-Board Diagnostics (OBD II) connector is located under the instrument panel on the driver's side.

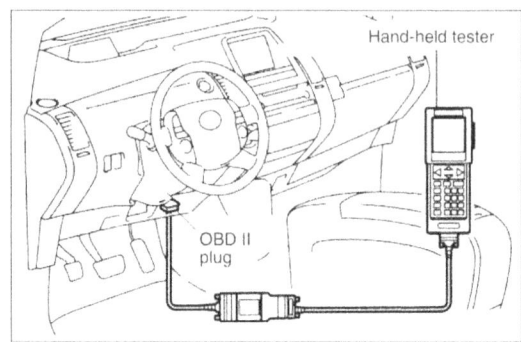

A large number of five-digit Diagnostic Trouble Codes (DTCs) exists for possible hybrid system faults. Some of these codes are Society of Automotive Engineers (SAE)-mandated generic DTCs. Other DTCs have definitions that are manufacturer-specific.

Toyota has added another level of diagnostics to the five-digit DTCs. Attached to many DTCs are three-digit Information Codes. These codes provide additional information and a more specific definition to the DTC. This helps narrow the diagnosis. Information Codes can only be read with the Techstream factory scan tool or equivalent.

No start condition: Be sure that the Service Plug is fully inserted and the Service Plug secondary latch is locked. If the interlock contacts do not complete their circuit, this will prevent hybrid system operation on 2004> Prius, Camry, and Altima hybrids.

SERVICE HINT: 	Verify that HV battery state-of-charge (SOC) is higher than 20%. If the vehicle has accident damage, be sure that there are no SRS or powertrain DTCs.
SERVICE HINT:	With a no-start condition, it is easy to confuse MG1 turning the engine over with a running engine. To verify, look at engine rpm on the scan tool. If the engine is not running, rpm will be lower than idle specification. Also, engine speed will not increase when the accelerator pedal is depressed, and there will likely be a DTC for failure of engine start.

Prius transmission PARK function: The transmission gearshift and PARK lock systems are electronic, and use the 12-volt auxiliary battery for power. If the auxiliary battery is discharged or disconnected, the vehicle cannot be started or shifted out of PARK. There is no mechanical linkage to the transmission.

Difficulty turning the Prius hybrid system OFF: If it is not possible to turn a Prius hybrid system OFF, there may be a malfunction in the shift control actuator which is preventing the transmission from engaging PARK. If this occurs, stop the vehicle and apply the parking brake. This bypasses the problem and permits shutdown.

SERVICE HINT: 	On 2004-2009 Prius models, extreme cold can prevent the combination meter from allowing the hybrid system to turn OFF. Toyota has issued a TSB for this condition.

Bladder fuel tank: 2001-2009 Prius models have a unique fuel tank with a flexible plastic bladder inside the steel fuel tank. Overfilling the fuel tank may force fuel into the EVAP system and into the EVAP system charcoal canister. Fuel soaked canisters may set DTCs and must be replaced.

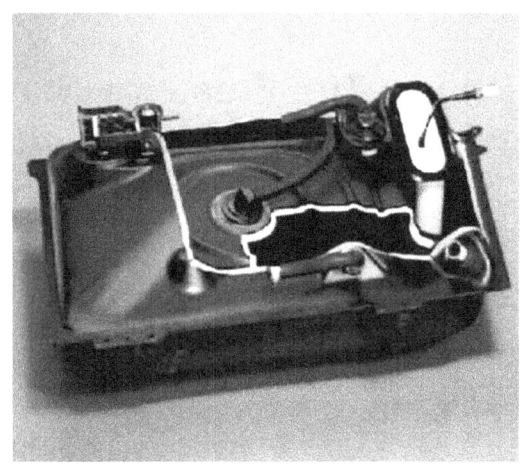

SERVICE HINT: 	Any problem with components associated with bladder style fuel tank will require fuel tank replacement (this includes fuel pump, fuel filter, fuel pressure regulator, fuel level sender, etc.).

SERVICE HINT:	Industry experience has shown that if fuel tank has been filled with diesel fuel, it is best to replace the fuel tank. The flexible plastic bladder may be damaged by the diesel fuel. Flushing the tank may temporarily resolve the condition, but the vehicle will likely return with DTCs.

Prius – Steering, setting zero point

Like other vehicles with electric power steering, the Prius steering column includes a steering torque sensor. The steering torque sensor zero point must be reset after any of the following occur:

- Whenever any front suspension or steering components are replaced.

- Whenever wheels are aligned.

- Whenever steering pulls to one side or steering effort right to left is different.

- If the power steering ECU is replaced.

SERVICE HINT: 	This procedure can only be performed with the Toyota Techstream scan tool or equivalent.

Prius – Tire Pressure Warning System (TPWS)

The Tire Pressure Warning System (TPWS) in the 2006> Prius uses individual tire pressure sensors in each tire. The sensor replaces the tire valve stem. In this Toyota system, each sensor has a unique ID number which must be recorded (registered) in the tire pressure warning ECU.

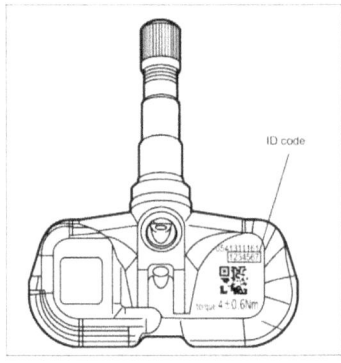

During the ID number registration process, the ID numbers for all four pressure sensors must be entered, even if only one sensor is replaced.

SERVICE HINT: 	Record all four TPWS sensor ID numbers before beginning the registration process. This procedure can only be performed with the Toyota Techstream scan tool or equivalent.

Appendix

Factory service information

Service information for hybrid vehicles is available to the general public, as well as trade professionals, from the manufacturers. Information is also available from third-party publishers. Here are links to a few currently-available internet sites. Be aware that internet locations, website addresses, and content are constantly changing.

Toyota Technical Information System (TIS):

http://www.techinfo.toyota.com

Nissan:

http://www.nissan-techinfo.com/

Ford:

http://www.helminc.com/helm/homepage.asp?r=http://www.helminc.com/helm/resource-center/ford-service-manuals-repair-information.htm

Honda:

http://www.helminc.com/helm/homepage.asp?r=http://www.helminc.com/helm/resource-center/ford-service-manuals-repair-information.htm

General Motors:

http://www.acdelcotechconnect.com/html/tss_tech_esi.jsp

Haynes Manuals: Toyota Prius 2001-2008

http://www.haynes.com/

Bentley Publishers: Toyota Prius Maintenance and Repair Manual 2004-2008

http://www.bentleypublishers.com/

Emergency Response Guides (ERGs)

Emergency Response Guides (ERGs, sometimes called First Responder Guides or FRGs) provide required valuable information to individuals who may respond to emergencies which may include hybrid vehicles. ERG information includes how to deal with the high voltage and chemical dangers inherent in hybrid vehicles.

Below are internet sites that provide downloadable ERGs and additional safety information. Be aware that internet locations, website addresses, and content are constantly changing.

The complete 2004-2009 Toyota Prius ERG is printed after the list below.

Toyota:

https://techinfo.toyota.com/techInfoPortal/appmanager/t3/ti;TISESSIONID=0R34J0k FB10KcHS4J2yhMJ7n0h1CWkzn7QTCylNgkRsbx63DcnpL!- 1292824609?_pageLabel=ti_erg&_nfpb=true

Nissan:

http://www.nissanusa.com/pdf/techpubs/altima_hybrid/2007/2007_Altima_Hybrid_F RG.pdf

Ford:

https://www.fleet.ford.com/showroom/2006fleetshowroom/pdfs/guide-escape.pdf

Honda:

https://techinfo.honda.com/rjanisis/pubs/web/RJAAI001_HYBRID.htm

GM

https://www.gmstc.com/FirstResponder.aspx

Third party sources:

http://www.firegraphics.org/ERG.htm

http://www.extrication.com/ERG.htm#Nissan_Altima_Hybrid_Vehicle_

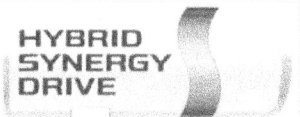
HYBRID SYNERGY DRIVE

2004 Model
2nd Generation

Emergency Response Guide

© 2004 Toyota Motor Corporation
All rights reserved. This document may not be
altered without the written permission of
Toyota Motor Corporation.

04PRIUSERG REV – (1/22/04)

Foreword

In May 2000, Toyota released the 1st generation Toyota Prius gasoline-electric hybrid vehicle in North America. Approximately 50,000 1st generation Prius were sold in the 2001 - 2003 model years. To educate and assist emergency responders in the safe handling of the 1st generation Prius hybrid technology, Toyota published the Prius Emergency Response Guide (M/N 00400-ERG02-0U).

With the release of the 2nd generation Prius in October 2003, this new 2004 model year Toyota Prius Emergency Response Guide was published for emergency responders. While many features from the 1st generation model are similar, emergency responders should recognize and understand the new, updated features of the 2nd generation Prius covered in this guide.

2nd Generation Prius New Features:

- Complete model change with a new exterior and interior design.
- Adoption of *Hybrid Synergy Drive* as the name for the Toyota Gasoline - Electric Hybrid System.
- *Hybrid Synergy Drive* includes a boost converter in the inverter assembly that boosts to 500-Volts the available voltage to the electric motor.
- The boost converter allows a reduction in the high voltage hybrid vehicle battery pack to 201-Volts.
- Addition of a high voltage 201-Volt motor driven air conditioning compressor.
- New electronic automatic transmission gearshift selector.
- Elimination of the conventional ignition switch with the new standard electronic key system and optional smart entry and start electronic key.
- Frontal airbags, optional side airbags for front occupants, and optional curtain shield airbags for front and rear occupants.

High voltage electrical safety remains an important factor in the emergency handling of the Prius *Hybrid Synergy Drive* system. It is important to recognize and understand the disabling procedures and warnings throughout the guide.

Additional topics contained in the guide include:

- Toyota Prius identification.
- Major *Hybrid Synergy Drive* component locations and descriptions.
- Extrication, fire, recovery, and additional emergency response information.
- Roadside assistance information.

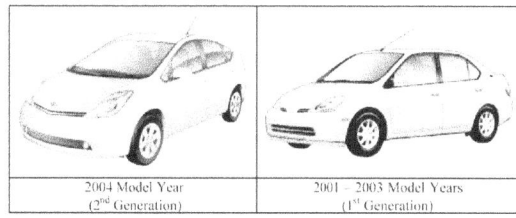

| 2004 Model Year (2nd Generation) | 2001 - 2003 Model Years (1st Generation) |

By following the information in this guide, emergency responders should be able to mitigate a rescue involving the 2nd generation Prius hybrid vehicle safely.

Note:
Emergency Response Guides for select Toyota alternative fuel vehicles may be viewed at *http://techinfo.toyota.com.*

-i-

Table of Contents	Page

About the Prius	1
Prius Identification	2
Hybrid Synergy Drive Component Locations & Descriptions	4
Electronic Key	6
Smart Entry & Start Electronic Key (Optional Equipment)	8
Electronic Gearshift Selector	10
Hybrid Synergy Drive Operation	11
Hybrid Vehicle (HV) Battery Pack and Auxiliary Battery	12
High Voltage Safety	13
SRS Airbags and Seat Belt Pretensioners	14
Emergency Response	15
Extrication	15
Fire	18
Overhaul	19
Recovery/Recycling NiMH HV Battery Pack	19
Spills	20
First Aid	20
Submersion	21
Roadside Assistance	22

About the Prius

The Toyota Prius continues into its 2nd generation as a gasoline-electric hybrid vehicle. The gasoline-electric hybrid system has been renamed *Hybrid Synergy Drive*. *Hybrid Synergy Drive* means the vehicle contains a gasoline engine and an electric motor for power. Two energy sources are stored on board the vehicle:

1. Gasoline stored in the fuel tank for the gasoline engine.
2. Electricity stored in a high voltage Hybrid Vehicle (HV) battery pack for the electric motor.

The result of combining these two power sources is increased fuel economy and reduced emissions. The gasoline engine also powers an electric generator to recharge the battery pack; unlike a pure all electric vehicle, the Prius never needs to be recharged from an external electric power source.

Depending on the driving conditions one or both sources are used to power the vehicle. The following illustration demonstrates how the Prius operates in various driving modes.

❶ On light acceleration at low speeds, the vehicle is powered by the electric motor. The gasoline engine is shut off.

❷ During normal driving the vehicle is powered mainly by the gasoline engine. The gasoline engine is also used to recharge the battery pack.

❸ During full acceleration, such as climbing a hill, both the gasoline engine and the electric motor power the vehicle.

❹ During deceleration, such as braking, the vehicle regenerates the kinetic energy from the front wheels to produce electricity that recharges the battery pack.

❺ While the vehicle is stopped, the gasoline engine and electric motor are off, however the vehicle remains on and operational.

-1-

Prius Identification

In appearance, the 2004 Prius is a 5-door hatchback. Exterior, interior, and engine compartment illustrations are provided to assist in identification.

The alphanumeric 17 character Vehicle Identification Number (VIN) is provided in the front windshield cowl and driver door post.

Example VIN: JTDKB20U840020208
(A Prius is identified by the first 6 alphanumeric characters **JTDKB2**)

Exterior
❶ **TOYOTA** *PRIUS* logos on rear hatchback door.
❷ Gasoline fuel filler door located on driver side rear quarter panel.

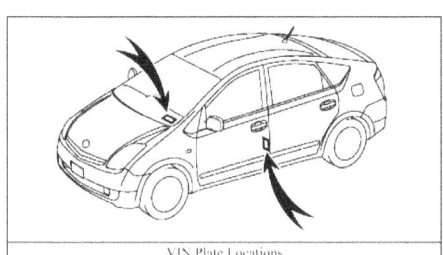
VIN Plate Locations

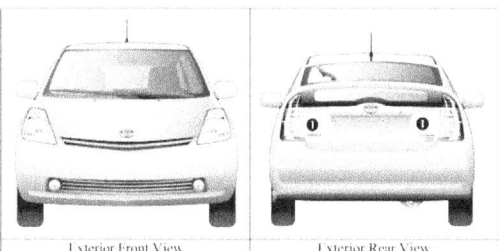
Exterior Front View   Exterior Rear View

Exterior Driver Side View

Exterior Rear and Driver Side View

-2-

Prius Identification (Continued)

Interior
❸ Dashboard mounted automatic transmission gearshift selector.
❹ Instrument cluster (speedometer, fuel gauge, **READY** light, warning lights) located in center dash and near the base of the windshield.
❺ LCD monitor (fuel consumption, energy monitor, radio controls, A/C controls) located above the center dash.

Engine Compartment
❻ 1.5 liter aluminum alloy gasoline engine.
❼ High voltage inverter/converter assembly with the logos on the cover.

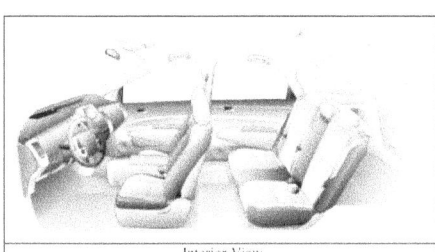
Interior View

Logos on Cover

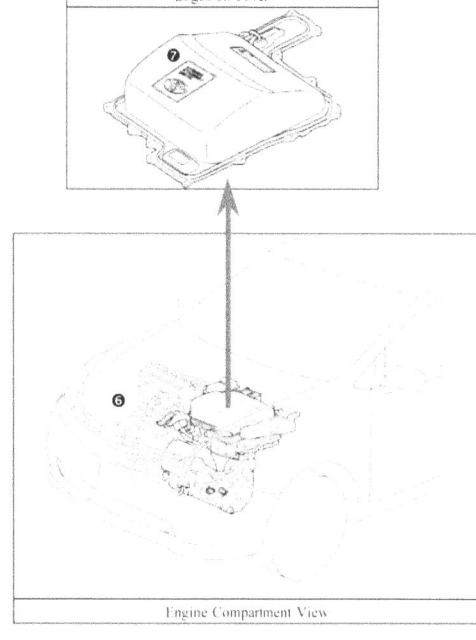
Engine Compartment View

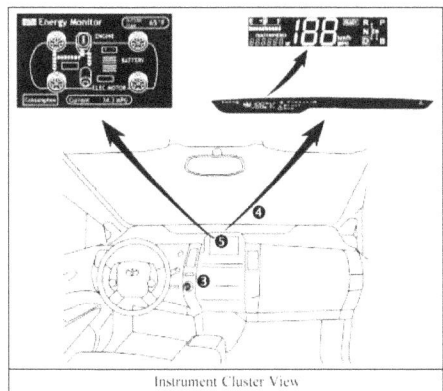
Instrument Cluster View

-3-

79

Hybrid Synergy Drive Component Locations & Descriptions

Component	Location	Description
12-Volt ❶ Auxiliary Battery	Cargo Area, Passenger Side	Low voltage lead-acid battery that controls all electrical equipment except electric motor, generator, inverter/converter, and A/C compressor.
Hybrid ❷ Vehicle (HV) Battery Pack	Cargo Area, Mounted to Cross Member and behind Rear Seat	201-Volt Nickel Metal Hydride (NiMH) battery pack consisting of 28 low voltage (7.2-volt) modules connected in series.
Power ❸ Cables	Under Carriage and Engine Compartment	Orange colored power cables carry high voltage Direct Current (DC) between the HV battery pack and inverter/converter. Also carries 3-phase Alternating Current (AC) between inverter/converter, motor, generator, and A/C compressor.
Inverter/ Converter ❹	Engine Compartment	Boosts and inverts the high voltage electricity from the HV battery pack to 3-phase AC electricity that drives the electric motor. The inverter/converter also converts AC electricity from the electric generator and motor (regenerative braking) to DC that recharges the HV battery pack.
Gasoline ❺ Engine	Engine Compartment	Provides two functions: 1) powers vehicle; 2) powers generator to recharge the HV battery pack. The engine is started and stopped under control of the vehicle computer.
Electric ❻ Motor	Engine Compartment	3-phase AC permanent magnetic electric motor contained in the transaxle. Used to power the vehicle.
Electric ❼ Generator	Engine Compartment	3-phase AC generator contained in the transaxle. Used to recharge the HV battery pack.
A/C ❽ Compressor	Engine Compartment	3-phase AC electrically driven motor compressor.
Fuel Tank ❾ and Fuel Lines	Undercarriage, Passenger Side	Fuel tank provides gasoline via a single fuel line to the engine. The fuel line is routed along passenger side under the floor pan.

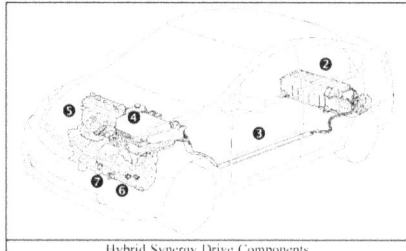

Hybrid Synergy Drive Components

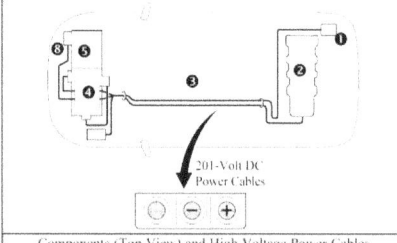

201-Volt DC Power Cables

Components (Top View) and High Voltage Power Cables

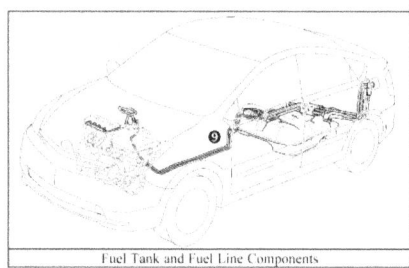

Fuel Tank and Fuel Line Components

-4-

Hybrid Synergy Drive Component Locations & Descriptions (Continued)

Key Specifications:

Gasoline Engine: 76 hp, 1.5 liter Aluminum Alloy Engine
Electric Motor: 67 hp, Permanent Magnet Motor
Transmission: Automatic Only
HV Battery: 201-Volt Sealed NiMH
Curb Weight: 2,890 lbs
Fuel Tank: 11.9 gals
Miles Per Gallon: 60/51 mpg (City/Hwy)
Liters/100 km: 4.0/4.2 L/100 km (City/Hwy)
Frame Material: Steel unibody
Body Material: Steel panels except aluminum hood and rear hatch.

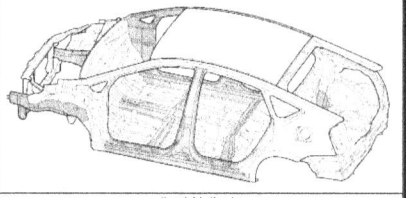

Steel Unibody

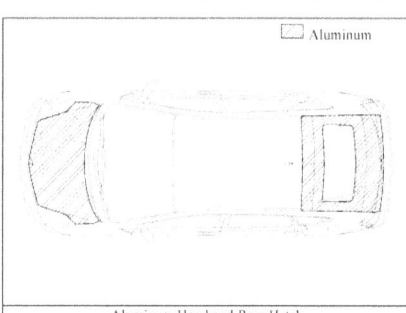

Aluminum

Aluminum Hood and Rear Hatch

-5-

80   Copyright © 2010 Automotive Aftermarket Training, Inc. All Rights Reserved.

Electronic Key

The 2004 Prius introduces a new electronic key as standard equipment.

Electronic key features:
- Wireless transmitter to lock/unlock the doors.
- Electronic key for starting.
- Hidden metal cut key to lock/unlock the doors from the driver exterior door lock.

Door (Lock/Unlock)
Two methods are available to lock/unlock the doors.

1. Pushing wireless electronic key lock/unlock buttons.

2. Inserting the hidden metal cut key in driver door lock and turning clockwise once unlocks the driver door; twice unlocks all doors. To lock all doors turn the key counter-clockwise once. Only the driver door contains an exterior door lock.

Vehicle Starting/Stopping
The electronic key has replaced the conventional metal cut key, and an electronic key slot and power button have replaced the ignition switch.

- A standard electronic key as shown in the illustration is inserted into the electronic key slot.

- The electronic key slot does not rotate like a conventional ignition switch. Instead, a power button with an integral status indicator light is provided above the electronic key slot to cycle through the various ignition modes. With the brake pedal released, the first push of the power button operates the accessory mode, the second push operates the ignition-on mode, and the third push turns the ignition off again.

Ignition Mode Sequence (Brake pedal released):

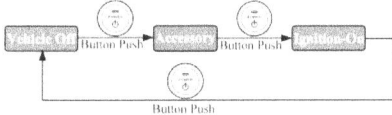

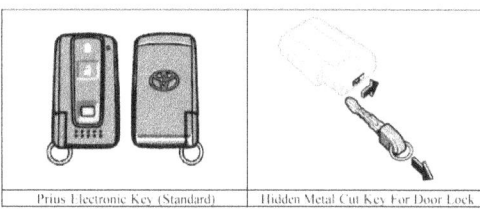

| Prius Electronic Key (Standard) | Hidden Metal Cut Key For Door Lock |

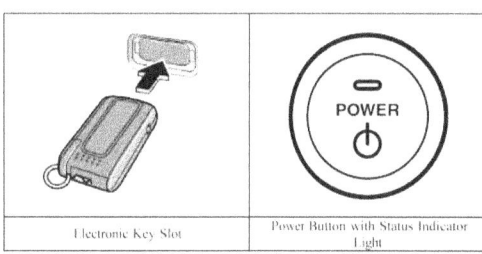

| Electronic Key Slot | Power Button with Status Indicator Light |

Ignition Mode	Power Button Indicator Light
Off	Off
Accessory	Green
Ignition-On	Amber
Vehicle Started (READY-On)	Off
Malfunction	Blinking Amber

-6-

Electronic Key (Continued)

Vehicle Starting/Stopping (Continued)

- Starting the vehicle takes priority over all other ignition modes and is accomplished by depressing the brake pedal and pushing the power button once. To verify the vehicle has started, the power button status indicator light is off and the **READY** light is illuminated in the instrument cluster.

- Once the vehicle has started and is on and operational (READY-on), the vehicle is shut off by bringing the vehicle to a complete stop and then depressing the power button once.

- The key slot prevents the electronic key from being removed while the vehicle is on and operational (READY-on) or in the ignition-on mode.

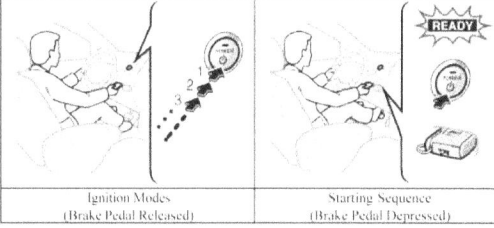

| Ignition Modes (Brake Pedal Released) | Starting Sequence (Brake Pedal Depressed) |

-7-

Smart Entry & Start Electronic Key (Optional Equipment)

The Prius may be equipped with an optional *smart entry and start electronic key* that appears similar in function and design to the standard electronic key. However, the smart key contains a transceiver that communicates bi-directionally enabling the vehicle to recognize the smart key in close proximity to the vehicle. The system can lock or unlock doors without pushing smart key buttons and start the hybrid system without inserting the smart key into the electronic key slot.

Smart key features:
- Passive (remote) function to lock/unlock the doors and start the vehicle.
- Wireless transmitter to lock/unlock the doors.
- Electronic key for starting.
- Hidden metal cut key to lock/unlock the doors from the driver door lock.

Door (Lock/Unlock)
Three methods are available to lock/unlock the doors.

1. Pushing wireless smart key lock/unlock buttons.

2. Touching the sensor on the backside of either exterior front door handle, with the smart key in close proximity to the vehicle, unlocks the doors. Pushing the black button on the front door handle locks the doors.

3. Inserting the metal cut key in driver door lock and turning clockwise once unlocks the driver door, twice unlocks all doors. To lock all doors turn the key counter-clockwise once. Only the driver door contains an exterior door lock.

Vehicle Starting/Stopping
The ignition modes and starting sequence are the same as the standard electronic key except the smart key does not have to be inserted into the electronic key slot.

- The optional smart key as shown in the illustrations may be inserted into the electronic key slot or kept in close proximity to the vehicle.

- With brake pedal released, the first push of the power button operates the accessory mode, the second push operates the ignition-on mode, and the third push turns the ignition off again.

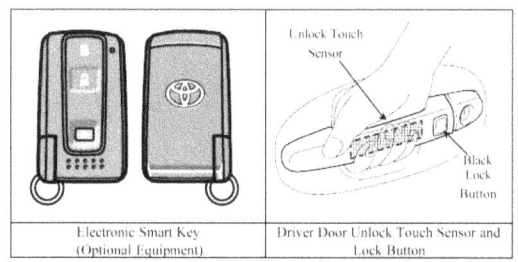

| Electronic Smart Key (Optional Equipment) | Driver Door Unlock Touch Sensor and Lock Button |

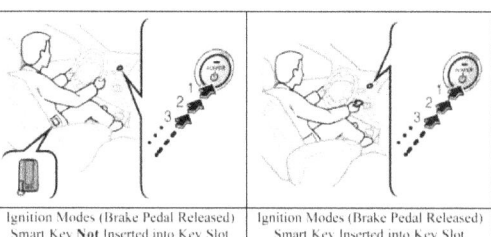

| Ignition Modes (Brake Pedal Released) Smart Key **Not** Inserted into Key Slot | Ignition Modes (Brake Pedal Released) Smart Key Inserted into Key Slot |

-8-

Smart Entry & Start Electronic Key (Optional Equipment) (Continued)

Vehicle Starting/Stopping (Continued)

Ignition Mode Sequence (Brake pedal released):

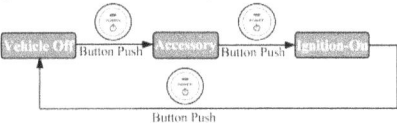

Ignition Mode	Power Button Indicator Light
Off	Off
Accessory	Green
Ignition-On	Amber
Vehicle Started (READY-On)	Off
Malfunction	Blinking Amber

- Starting the vehicle takes priority over all other ignition modes and is accomplished by depressing the brake pedal and pushing the power button once. To verify the vehicle has started, the power button status indicator light is off and the **READY** light is illuminated in the instrument cluster.

- Once the vehicle has started and is on and operational (READY-on), the vehicle is shut off by bringing the vehicle to a complete stop and then depressing the power button once.

- Vehicles equipped with the optional smart key have a disabling button located beneath the steering column as shown in the illustration. When disabled, the smart key must be inserted into the key slot to enable the ignition modes or start the vehicle.

- The key slot prevents the electronic key from being removed while the vehicle is on and operational (READY-on) or in the ignition-on mode.

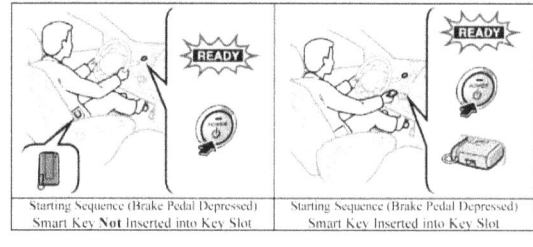

| Starting Sequence (Brake Pedal Depressed) Smart Key **Not** Inserted into Key Slot | Starting Sequence (Brake Pedal Depressed) Smart Key Inserted into Key Slot |

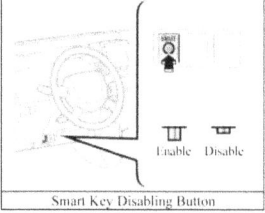

| Smart Key Disabling Button |

-9-

Electronic Gearshift Selector

The Prius electronic gearshift selector is a newly developed momentary select shift-by-wire system that engages the transaxle in **R**everse, **N**eutral, **D**rive, or engine **B**rake modes.

- These modes may only be engaged while the vehicle is on and operational (READY-on), except for **N**eutral which may also be engaged while in the ignition-on mode. After selecting the gear position R, N, D, or B the transaxle remains in that position, identified on the instrument cluster, but the shift selector returns to a default position.

- Unlike a conventional vehicle, the electronic shift selector does not contain a park position. Instead, a separate **P** switch located above the shift selector engages the park position.

- When the vehicle is stopped, regardless of shift selector position, the electro-mechanical parking pawl is engaged to lock the transaxle into park by either depressing the **P** switch or pushing the power button to shut off the vehicle.

- Being electronic, the gearshift selector and the park systems depend on the low voltage 12-Volt auxiliary battery for power. If the 12-Volt auxiliary battery is discharged or disconnected, the vehicle cannot be started and cannot be shifted out of park.

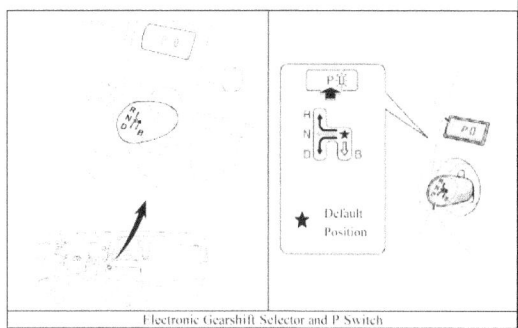

Electronic Gearshift Selector and P Switch

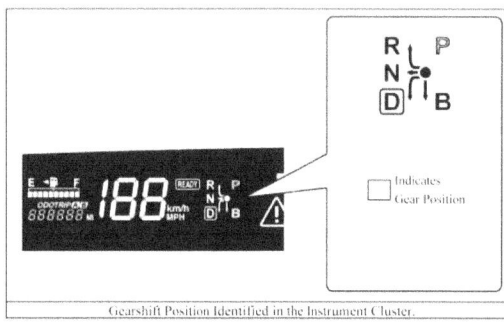

Gearshift Position Identified in the Instrument Cluster.

-10-

Hybrid Synergy Drive Operation

Once the **READY** indicator is illuminated in the instrument cluster, the vehicle may be driven. However, the gasoline engine does not idle like a typical automobile and will start and stop automatically. It is important to recognize and understand the **READY** indicator provided in the instrument cluster. When lit, it informs the driver the vehicle is on and operational even though the gasoline engine may be off and the engine compartment is silent.

Vehicle Operation
- With the Prius, the gasoline engine may stop and start at any time while the **READY** indicator is on.

- Never assume the vehicle is shut off just because the engine is off. Always look for the **READY** indicator status. The vehicle is shut off when the **READY** indicator is off.

- The vehicle may be powered by:
    1. The electric motor only.
    2. The gasoline engine only.
    3. A combination of both the electric motor and the gasoline engine.

- The vehicle computer determines the mode in which the vehicle operates to improve fuel economy and reduce emissions. The driver cannot manually select the mode.

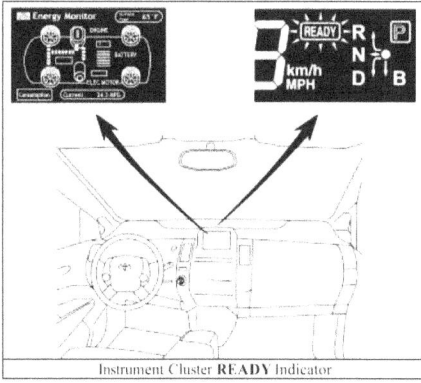

Instrument Cluster **READY** Indicator

-11-

Hybrid Vehicle (HV) Battery Pack and Auxiliary Battery

The Prius contains a high voltage, Hybrid Vehicle (HV) battery pack and a low voltage auxiliary battery. The HV battery pack contains non-spillable, sealed Nickel Metal Hydride (NiMH) battery modules and the auxiliary battery is a typical automotive lead-acid type.

HV Battery Pack

- The HV battery pack is enclosed in a metal case and is rigidly mounted to the cargo area floor pan cross member behind the rear seat. The metal case is isolated from high voltage and concealed by a cover in the cargo area.

- The HV battery pack consists of 28 low voltage (7.2-Volt) NiMH battery modules connected in series to produce approximately 201-Volts. Each NiMH battery module is non-spillable and sealed in a plastic case.

- The electrolyte used in the NiMH battery module is an alkaline of potassium and sodium hydroxide. The electrolyte is absorbed into the battery cell plates and will form a gel that will not normally leak, even in a collision.

- In the unlikely event the battery pack is overcharged, the modules vent gases directly outside the vehicle through a vent hose connected to each NiMH battery module.

HV Battery Pack	
Battery pack voltage	201-Volts
Number of NiMH battery modules in the pack	28
Battery pack weight	86 lbs/39 kg
NiMH battery module voltage	7.2-Volts
NiMH battery module dimensions	11 x 3/4 x 4 inches 27.9 x 1.9 x 10.1 cm
NiMH battery module weight	2.2 lbs/1 kg

Components Powered by the HV Battery Pack

- Electric Motor
- Inverter/Converter
- A/C Compressor
- Electric Generator
- Power Cables

HV Battery Pack Recycling

- The HV battery pack is recyclable. Contact the nearest Toyota dealer, or:
  United States: (800) 331-4331
  Canada: (888) Toyota 8 [(888)-869-6828]

Auxiliary Battery

- The Prius also contains a lead-acid 12-Volt battery. This 12-Volt auxiliary battery powers the vehicle electrical system similar to a conventional vehicle. As with other conventional vehicles, the auxiliary battery is grounded to the metal chassis of the vehicle.

- The auxiliary battery is located in the passenger side rear cargo area. It also contains a hose to vent gases outside the vehicle if overcharged.

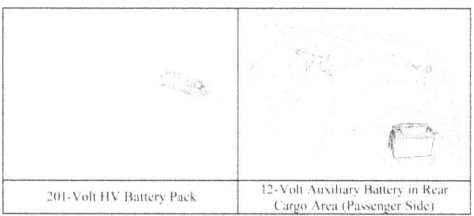

| 201-Volt HV Battery Pack | 12-Volt Auxiliary Battery in Rear Cargo Area (Passenger Side) |

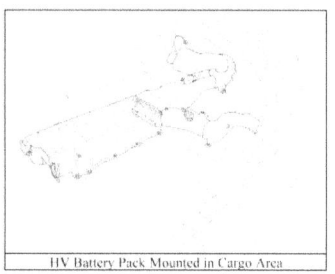

HV Battery Pack Mounted in Cargo Area

-12-

High Voltage Safety

The HV battery pack powers the high voltage electrical system with DC electricity. Positive and negative high voltage power cables are routed from the battery pack, under the vehicle floor pan, to the inverter/converter. The inverter/converter contains a circuit that boosts the HV battery voltage from 201 to 500-Volts DC. The inverter creates 3-phase AC to power the motors in the engine compartment. Sets of 3 power cables are routed from the inverter to each high voltage motor (electric motor, electric generator, and A/C compressor). Occupants in the vehicle and emergency responders are separated from high voltage electricity by the following systems:

High Voltage Safety System

- A high voltage fuse ❶ provides short circuit protection in the HV battery pack.

- Positive and negative high voltage power cables ❷ connected to the HV battery pack are controlled by 12-Volt normally open relays ❸. When the vehicle is shut off, the relays stop electricity flow from the HV battery pack.

> **WARNING:**
> - *Power remains in the high voltage electrical system for 5 minutes after the HV battery pack is shut off.*
> - ***Never*** *touch, cut, or open any orange high voltage power cable or high voltage component.*

- Both positive and negative power cables ❷ are isolated from the metal chassis, so there is no possibility of shock by touching the metal chassis.

- A ground fault monitor ❹ continuously monitors for high voltage leakage to the metal chassis while the vehicle is running. If a malfunction is detected, the vehicle computer ❺ will illuminate the master warning light ⚠ in the instrument cluster and the hybrid warning light 🔧 in the LCD display.

- The HV battery pack relays will automatically open to stop electricity flow in a collision sufficient to activate the SRS airbags.

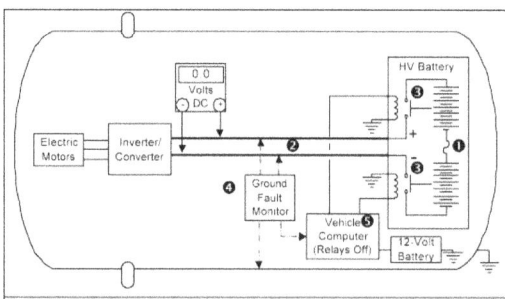

High Voltage Safety System – Vehicle Shut Off (READY-off)

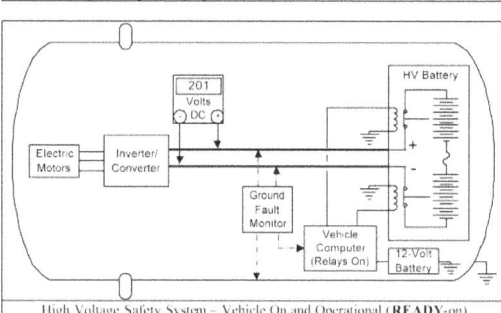

High Voltage Safety System – Vehicle On and Operational (READY-on)

-13-

SRS Airbags and Seat Belt Pretensioners

Standard Equipment
- Electronic frontal impact sensors (2) are mounted in the engine compartment ❶.
- Front seat belt pretensioners are mounted near the base of the B-pillar ❷.
- Frontal dual stage airbag for the driver ❸ is mounted in the steering wheel hub.
- Frontal dual stage airbag for the front passenger ❹ is integrated into the dashboard and deploys through the top of the dashboard.
- SRS computer ❺ is mounted on the floor pan underneath the center console. It also contains an impact sensor.

Optional Side Impact Airbag Package
- Front electronic side impact sensors (2) are mounted near the base of the B-pillars ❻.
- Rear electronic side impact sensors (2) are mounted near the base of the C-pillars ❼.
- Front seat side impact airbags ❽ are mounted in the front seats.
- Curtain shield side impact airbags ❾ are mounted along the outer edge inside the roof rails.

WARNING:
- *The SRS computer is equipped with a back up source that powers the SRS airbags up to **90 seconds** after disabling the vehicle.*
- *The front seat side airbags and the curtain shield side airbags may deploy independent of each other.*

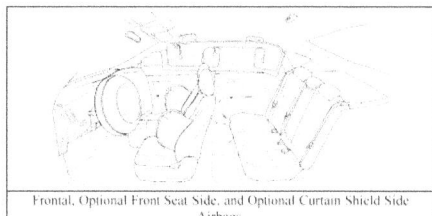

Frontal, Optional Front Seat Side, and Optional Curtain Shield Side Airbags.

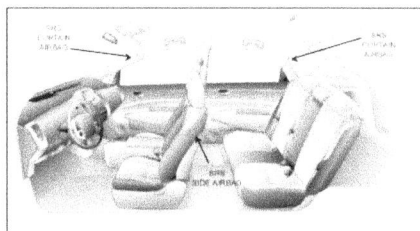

Front Seat and Curtain Shield Side Airbag Identifiers

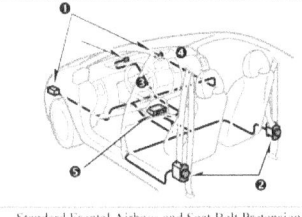

Standard Frontal Airbags and Seat Belt Pretensioners

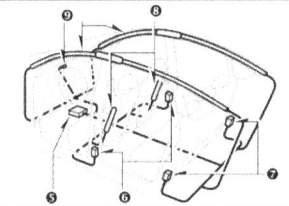

Optional Front Seat and Curtain Shield Side Airbags

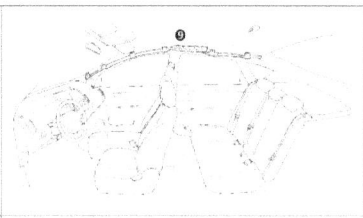

Curtain Shield Side Airbag Inflator in Roof Rail

-14-

Emergency Response

On arrival, emergency responders should follow their standard operating procedures for vehicle incidents. Emergencies involving the Prius may be handled like other automobiles except as noted in these guidelines for Extrication, Fire, Overhaul, Recovery, Spills, First Aid, and Submersion.

WARNING:
- ***Never** assume the Prius is shut off simply because it is silent.*
- *Always observe the instrument cluster for the **READY** indicator status to verify whether the vehicle is on or shut off.*

Extrication
- Immobilize Vehicle
  Chock wheels and set the parking brake.
  Push the **P** switch to engage park.

- Disable Vehicle
  Performing either of the two procedures will shut the vehicle off and disable the HV battery pack, SRS airbags, and gasoline fuel pump.

  Procedure #1
  1. Confirm the status of **READY** indicator in the instrument cluster.
  2. If the **READY** indicator is illuminated, the vehicle is on and operational. Shut off the vehicle by pushing the power button once.
  3. The vehicle is already shut off if the instrument cluster lights and the **READY** indicator are **not** illuminated. Do **not** push the power button the vehicle may start.
  4. Remove the electronic key from the key slot.
  5. If equipped, disable the smart key button underneath the steering column.
  6. Keep the electronic key at least 16 feet (5 meters) away from the vehicle.
  7. If the electronic key cannot be removed from the key slot or if the electronic key cannot be found, disconnect the 12-Volt auxiliary battery in the rear cargo area.

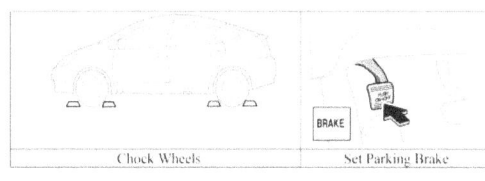

Chock Wheels | Set Parking Brake

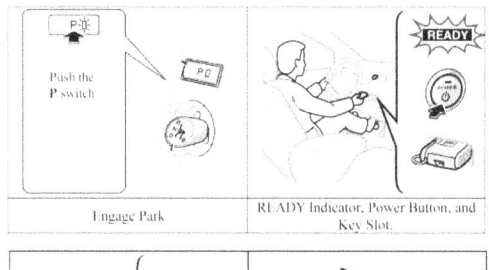

Engage Park | READY Indicator, Power Button, and Key Slot.

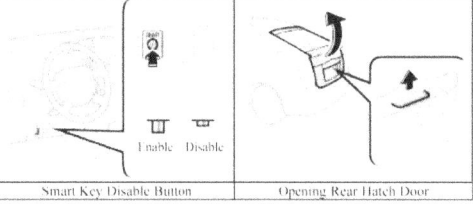

Smart Key Disable Button | Opening Rear Hatch Door

-15-

Emergency Response (Continued)

Extrication (Continued)

Alternate Procedure (power button inaccessible)

Procedure #2
1. Disconnect the 12-Volt auxiliary battery in the rear cargo area.
2. Remove the HEV fuse (20A yellow colored) in the engine compartment junction block as illustrated. When in doubt, pull all four fuses in the fuse block.

WARNING:
- *After disabling the vehicle, power is maintained for **90 seconds** in the SRS system and **5 minutes** in the high voltage electrical system.*
- *If either of the disabling procedures above cannot be performed, proceed with caution as there is no assurance that the high voltage electrical system, SRS, or fuel pump are disabled.*
- ***Never*** *touch, cut, or open any orange high voltage power cable or high voltage component.*

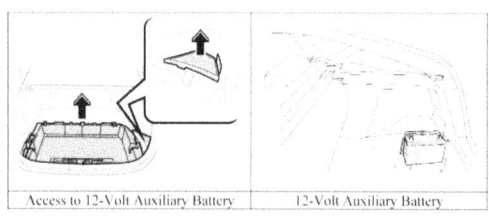

| Access to 12-Volt Auxiliary Battery | 12-Volt Auxiliary Battery |

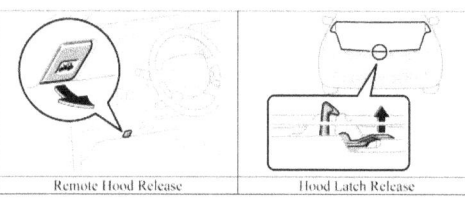

| Remote Hood Release | Hood Latch Release |

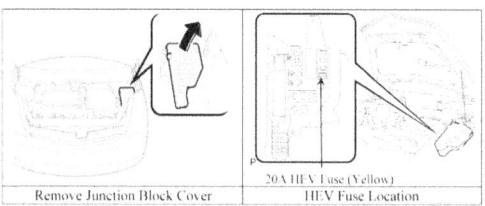

| Remove Junction Block Cover | 20A HEV Fuse (Yellow) / HEV Fuse Location |

-16-

Emergency Response (Continued)

Extrication (Continued)

- **Stabilize Vehicle**
  Crib at (4) points directly under the front and rear pillars.
  Do not place cribbing under the high voltage power cables, exhaust system, or fuel system.

- **Access Patients**
  Glass Removal
  Use normal glass removal procedures as required.

  SRS Awareness
  Responders need to be cautious when working in close proximity to undeployed airbags and seat belt pretensioners. Deployed front dual stage airbags automatically ignite both stages within a fraction of a second.

  Door Removal/Displacement
  Doors can be removed by conventional rescue tools such as hand, electric, and hydraulic. In certain situations, it may be easier to pry back the body to expose and unbolt the hinges.

  Roof Removal
  The vehicle may contain optional curtain shield airbags. If equipped and undeployed, it is not recommend to remove or to displace the roof. Optional curtain shield airbags may be identified as illustrated.

  Dash Displacement
  The vehicle may contain optional curtain shield airbags. When equipped, do not remove or displace the roof during a dash displacement to avoid cutting into the airbags or inflators. As an alternative, dash displacement may be performed by using a Modified Dash Roll.

  If not equipped with the optional curtain shield airbags, displace the dash by using a conventional dash roll, Modified Dash Roll, or jacking the dash.

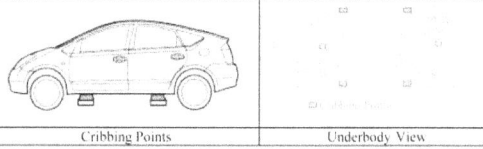

| Cribbing Points | Underbody View |

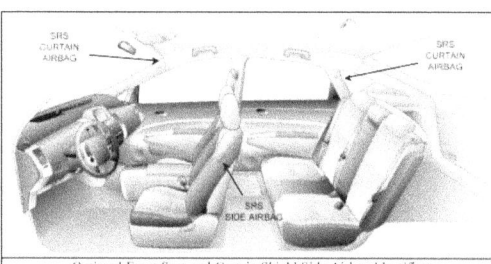

| Optional Front Seat and Curtain Shield Side Airbag Identifiers |

-17-

Emergency Response (Continued)

Extrication (Continued)

Rescue Lift Air Bags
Responders should not place cribbing or rescue lift airbags under the high voltage power cables, exhaust system, or fuel system.

Repositioning Steering Wheel and Seat
Tilt steering and seat controls are shown in the illustration

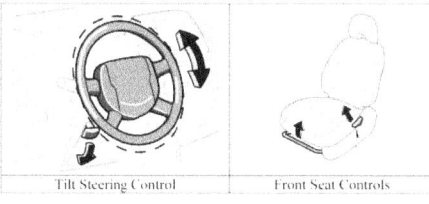

| Tilt Steering Control | Front Seat Controls |

Fire
Approach and extinguish a fire using proper vehicle fire fighting practices as recommended by NFPA, IFSTA, or the National Fire Academy (USA).

- Extinguishing Agent
Water has been proven to be a suitable extinguishing agent.

- Initial Fire Attack
Perform a fast, aggressive fire attack.
Divert the runoff from entering watershed areas.

Attack teams may not be able to identify a Prius until the fire has been knocked down and overhaul operations have commenced.

- Fire in the HV Battery Pack
Should a fire occur in the NiMH HV battery pack, the incident commander will have to decide whether to pursue an offensive or defensive attack.

> **WARNING:**
> - *Potassium hydroxide and sodium hydroxide are key ingredients in the NiMH battery module electrolyte.*
> - *The modules are contained within a metal case and access is limited to a small opening on the top.*
> - *The cover should **Never** be breached or removed under any circumstances, including fire. Doing so may result in severe electrical burns, shock or electrocution.*

-18-

Emergency Response (Continued)

Fire (Continued)

When allowed to burn themselves out, the Prius NiMH battery modules burn rapidly and can quickly be reduced to ashes except for the metal alloy cell plates.

Offensive Fire Attack
Flooding the HV battery pack, located in the cargo area, with copious amounts of water at a safe distance will effectively control the HV battery pack fire by cooling the adjacent NiMH battery modules to a point below their ignition temperature. The remaining modules on fire, if not extinguished by the water, will burn themselves out.

Defensive Fire Attack
If the decision has been made to fight the fire using a defensive attack, the fire attack crew should pull back a safe distance and allow the NiMH battery modules to burn themselves out. During this defensive operation, fire crews may utilize a water stream or fog pattern to protect exposures or to control the path of smoke.

Overhaul
During overhaul, if not already done, immobilize and disable the vehicle. See illustrations on page 15.

- Immobilize Vehicle
Chock wheels and set the parking brake.
Push the **P** switch to engage park.

- Disable Vehicle
Performing either of the two procedures will shut the vehicle off and disable the HV battery pack, SRS airbags, and gasoline fuel pump.

Procedure #1
1. Confirm the status of **READY** indicator in the instrument cluster.
2. If the **READY** indicator is illuminated, the vehicle is on and operational. Shut off the vehicle by pushing the power button once.

3. The vehicle is already shut off if the instrument cluster lights and the **READY** indicator are **not** illuminated. Do **not** push the power button the vehicle may start.
4. Remove the electronic key from the key slot.
5. If equipped, disable the smart key button underneath the steering column.
6. Keep the electronic key at least 16 feet (5 meters) away from the vehicle.
7. If the electronic key cannot be removed from the key slot or if the electronic key cannot be found, disconnect the 12-Volt auxiliary battery in the rear cargo area.

Alternate Procedure (power button inaccessible)

Procedure #2
1. Disconnect the 12-Volt auxiliary battery in the rear cargo area.
2. Remove the HEV fuse (20A yellow colored) in the engine compartment junction block as illustrated on page 16. When in doubt, pull all four fuses in the fuse block.

> **WARNING:**
> - *After disabling the vehicle, power is maintained for **90 seconds** in the SRS system and **5 minutes** in the high voltage electrical system.*
> - *If either of the disabling steps above cannot be performed, proceed with caution as there is no assurance that the high voltage electrical system, SRS, or fuel pump are disabled.*
> - ***Never** touch, cut, or open any orange high voltage power cable or high voltage component.*

Recovery/Recycling NiMH HV Battery Pack
Clean up of the HV battery pack can be accomplished by the vehicle recovery crew without further concern from runoff or spill. For information regarding recycling of the HV battery pack, contact the nearest Toyota dealer, or:

United States: (800) 331-4331
Canada: (888) Toyota 8 [(888)-869-6828]

-19-

Emergency Response (Continued)

Spills

The Prius contains the same common automotive fluids used in other Toyota vehicles, with the exception of NiMH electrolyte used in the HV battery pack. The NiMH battery electrolyte is a caustic alkaline (pH 13.5) that is damaging to human tissues. The electrolyte, however, is absorbed in the cell plates and will not normally spill or leak out even if a battery module is cracked. A catastrophic crash that would breach both the metal battery pack case and the plastic battery module would be a rare occurrence.

Similar to using baking soda to neutralize a lead-acid battery electrolyte spill, a dilute boric acid solution or vinegar is used to neutralize a NiMH battery electrolyte spill.

During an emergency, Toyota Material Safety Data Sheets (MSDS) may be requested by contacting:

> United States: CHEMTREC at (800) 424-9300
> Canada: CANUTEC at *666 or (613) 996-6666 (collect)

- Handle NiMH Electrolyte Spills Using The Following Personal Protective Equipment (PPE):
  Splash shield or safety goggles. Fold down helmet shields are not acceptable for acid or electrolyte spills.
  Rubber, latex or Nitrile gloves.
  Apron suitable for alkaline.
  Rubber boots.

- Neutralize NiMH Electrolyte
  Use a boric acid solution or vinegar.
  Boric acid solution - 800 grams boric acid to 20 liters water or 5.5 ounces boric acid to 1 gallon of water.

First Aid

Emergency responders may not be familiar with a NiMH electrolyte exposure when rendering aid to a patient. Exposure to the electrolyte is unlikely except in a catastrophic crash or through improper handling. Utilize the following guidelines during an exposure.

WARNING:
The NiMH battery electrolyte is a caustic alkaline (pH 13.5) that is damaging to human tissue.

- Wear Personal Protective Equipment (PPE)
  Splash shield or safety goggles. Fold down helmet shields are not acceptable for acid or electrolyte spills.
  Rubber, latex or Nitrile gloves.
  Apron suitable for alkaline.
  Rubber boots.

- Absorption
  Perform gross decontamination by removing affected clothing and properly disposing of the garments.
  Rinse the affected areas with water for 20 minutes.
  Transport to the nearest emergency medical care facility.

- Inhalation Non-Fire Situations
  No toxic gases are emitted under normal conditions.

- Inhalation Fire Situations
  Toxic gases are given off as the by-product of combustion. All responders in the Hot Zone should wear the proper PPE for fire fighting including SCBA.
  Remove patient from the hazardous environment to a safe area and administer oxygen.
  Transport to the nearest emergency medical care facility.

- Ingestion
  Do not induce vomiting.
  Allow patient to drink large quantities of water to dilute electrolyte (Never give water to an unconscious person).
  If vomiting occurs spontaneously, keep patients head lowered and forward to reduce the risk of aspiration.
  Transport to the nearest emergency medical care facility.

-20-

Emergency Response (Continued)

Submersion

Handle a Prius that is fully or partially submerged in water by disabling the HV battery pack, SRS airbags, and gasoline fuel pump.

- Remove vehicle from the water.

- Drain water from the vehicle if possible.

- Follow the immobilizing and disabling procedures on page 15.

Roadside Assistance

The Prius utilizes an electronic gearshift selector and an electronic **P** switch for park. If the 12-Volt auxiliary battery is discharged or disconnected, the vehicle cannot be started nor can it be shifted out of park. If discharged, the 12-Volt auxiliary battery can be jump started to allow vehicle starting and shifting out of park. Most other roadside assistance operations may be handled like conventional Toyota vehicles.

Toyota Roadside Assistance is available during the basic warranty period by contacting:

   United States: (877) 304-6495
   Canada: (888) TOYOTA 8   [(888) 869-6828]

Towing
The Prius is a front wheel drive vehicle and it **must** be towed with the front wheels off the ground. Failure to do so may cause serious damage to Hybrid Synergy Drive components.

Vehicle Operation
Refer to the Electronic Key section page 6 for vehicle starting/stopping and page 15 for vehicle disabling information.

- The vehicle may be shifted out of park into **Neutral** only in the ignition-on and READY-on modes.

- If the 12-Volt auxiliary battery is discharged, the vehicle will not start and shifting out of park is not possible. There is no manual override except to jump start the vehicle.

Spare Tire
The spare tire, jack, and tools are provided in the cargo area as illustrated. The spare tire is for temporary use only (do not exceed 50 mph/80 kph).

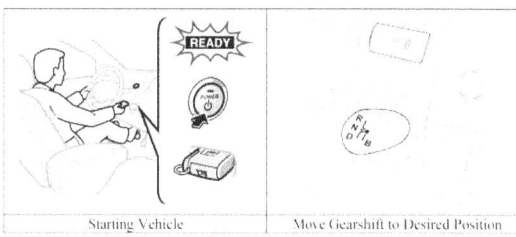

Starting Vehicle | Move Gearshift to Desired Position

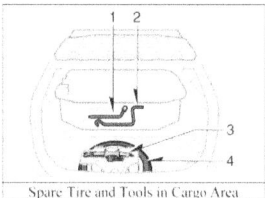

Spare Tire and Tools in Cargo Area

-22-

Roadside Assistance (Continued)

Jump Starting
The 12-Volt auxiliary battery may be jump started if the vehicle does not start and the instrument cluster gauges are dim or off after depressing the brake pedal and pushing the power button.

The 12-Volt auxiliary battery is located in the cargo area. The rear hatch door will not unlock or open if the auxiliary battery is discharged. Instead, an accessible remote 12-Volt auxiliary battery positive terminal is provided in the engine compartment junction block, as illustrated, for jump starting.

- Remove the junction block cover and connect the positive jumper cable to the positive terminal in the junction block.

- Connect the negative terminal to the ground nut.

- The high voltage HV battery pack cannot be jump started.

Immobilizer & Anti-Theft Alarm
The vehicle comes standard with an electronic key immobilizer system. An anti-theft alarm is optional equipment.

- The vehicle may only be started with a learned immobilizer coded electronic key.

- To disable the optional alarm use the unlock button on the electronic key, unlock the driver door with the hidden metal cut key, or engage the ignition-on mode.

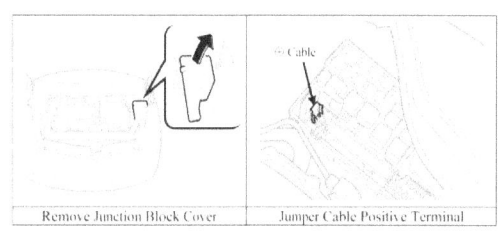

Remove Junction Block Cover | Jumper Cable Positive Terminal

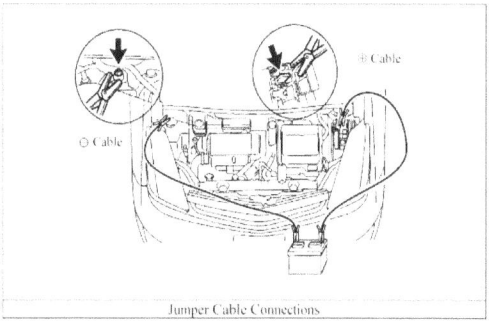

Jumper Cable Connections

-23-

www.ingramcontent.com/pod-product-compliance
Lightning Source LLC
Chambersburg PA
CBHW081143170526
45165CB00008B/2782